Universitext

George R. Kempf

Complex Abelian Varieties and Theta Functions

Springer-Verlag

Berlin Heidelberg New York
London Paris Tokyo
Hong Kong Barcelona

George R. Kempf
Department of Mathematics
John Hopkins University
Baltimore, MD 21218, USA

Mathematics Subject Classification (1980):
14 K 20, 14 K 25, 32 C 35, 32 J 25, 32 N 05

ISBN-13: 978-3-540-53168-5   e-ISBN-13: 978-3-642-76079-2
DOI: 10.1007/978-3-642-76079-2

Library of Congress Cataloging-in-Publication Data
Kempf, George
Complex abelian varieties and theta functions / George R. Kempf. p. cm. –
(Universitext). Includes bibliographical references and index.
1. Abelian varieties. 2. Functions, Theta. I. Title
QA564.K45 1990
516.3'53 – dc20 90-22573 CIP

41/3140–543210 – Printed on acid-free paper

Preface

The study of abelian varieties began with the one-dimensional case of elliptic curves. As such curves are defined by a general cubic polynomial equation in two variables, their study is basic to all but the simplest mathematics. The modern approach to elliptic curves occurred in the beginning of the nineteen century with the work of Gauss, Abel and Jacobi.

Since the classical period there have been many developments in mathematics. There are basically two distinct lines of generalization of an elliptic curve. They are algebraic curves of higher genus > 1. The other is higher dimensional compact algebraic groups (abelian varieties).

This book deals with these higher dimensional objects which surprisingly enough have more similar properties to elliptic curves than curves of higher genus. There are three methods for studying abelian varieties: arithmetic, algebraic and analytic. The arithmetic study properly using both the algebraic and analytic approaches and reduction modulo a prime. Mumford's book [3] presents an adequate introduction to the algebraic approach with some indication of the analytic theory. In this book I have restricted attention to the analytic approach and I try to make full use of complex Hermitian geometry.

In this book I give the basic material on abelian varieties, their invertible sheaves and sections, and cohomology and associated mappings to projective spaces. I also provide an introduction to the moduli (parameter spaces for abelian varieties) and modular functions. Lastly I give some examples where abelian varieties occur in mathematics.

Some of the material is parallel to that found in Igusa's book [1], but I have tried to develop the subject geometrically and avoid the connection with representations of infinite non-abelian groups in Hilbert space. The book brings some developments from the literature to book form; for example, Mumford's theory of the theta group acting on the space of sections of invertible sheaves.

It seems an impossible task to give a proper bibliography and history of the last two hundred years. As we desire, one generation's theorems have become examples of the next generation's theories. For the people (some very famous) we don't mention explicitly who have participated in the historical development of this branch of mathematics, we give thanks for their efforts.

Baltimore, August 1990   *George R. Kempf*

Table of Contents

Chapter 1. Complex Tori

§ 1.1 The Definition of Complex Tori

The lattice L in a real or complex finite dimensional vector space V is a discrete subgroup such that the quotient group V/L is compact. The lattice L is a free Abelian group of rank equal to the real dimension of V and the induced mapping $L \otimes_{\mathbb{Z}} \mathbb{R} \to V$ is an isomorphism and conversely.

A complex torus $X = V/L$ is a complex vector space V modulo a lattice L. Thus a complex torus is a commutative compact complex $\mathbb{C}$-analytic group. The complex tangent space $\mathrm{Lie}(X)$ of X at the identify 0 is naturally identified with V. The quotient homomorphism $V \to V/L$ is just the exponential mapping. $\exp : \mathrm{Lie}(X) \to X$.

This explicit picture of complex tori is complemented by an abstract characterization (for those who know the elements of complex Lie group theory).

Theorem 1.1. *Any compact connected $\mathbb{C}$-analytic group X is a complex torus.*

Proof. First we assume that X is known to be commutative. Then the exponential $\exp : \mathrm{Lie}(X) \to X$ will be a $\mathbb{C}$-analytic homomorphism which is locally an isomorphism. The last property implies that the kernel L of $\exp$ is discrete. As X is connected $\exp$ is surjective and, hence, $X \approx \mathrm{Lie}(X)/L$. Thus L is a lattice as X is compact.

To see that X is commutative consider the adjoint representation $\mathrm{Ad} : X \to \mathrm{Aut}(\mathrm{Lie}(X))$. This is a $\mathbb{C}$-analytic mapping from a compact variety to an affine variety. Thus $\mathrm{Ad}(X) = \mathrm{Identity}$. Hence $T_0 \mathrm{Ad} = \mathrm{ad} : \mathrm{Lie}(X) \to \mathrm{End}(\mathrm{Lie}(X))$ is zero. In other words $\mathrm{Lie}(X)$ is an abelian Lie algebra. Consequently the connected group X is commutative. $\qquad\square$

To get a rough idea of the possible complex tori V/L, we assume that $l_1, \ldots, l_g, m_1, \ldots, m_g$ is a basis for L and that $l_1, \ldots, l_g$ are a complex basis of $\mathbb{C}$ (we may always make such a choice). Hence $g = \dim_{\mathbb{C}}(V)$. Then $m_i = \sum_{1 \le j \le g} \alpha_{i,j} l_j$ where $(\alpha_{i,j})$ is a $g \times g$ complex matrix. The condition that we have a lattice is that $l_1, \ldots, l_g, m_1, \ldots, m_g$ are $\mathbb{R}$-linearly independent or, what is the same, $\mathrm{Im}(\alpha_{i,j})$ is an invertible matrix. The abstract moral of this is that the space of complex tori with a properly marked basis is naturally an analytic manifold of $\dim g^2$ (in this case the $\alpha_{i,j}$ are global coordinates on these spaces).

Exercise 1. Let X be a complex torus of dimension g. Then for any non-zero integer on multiplication by $m : X \to X$ is a surjective homomorphism with finite kernel X_m and $\#X_m = |m|^{2g}$.

Exercise 2. Let $\phi : V/L \to V'/L'$ be a homomorphism of $\mathbb{C}$-analytic groups between two complex tori. Show that there is a unique $\mathbb{C}$-linear mapping $A : V \to V'$ such that $A(L) \subset L'$ which induces ϕ.

Exercise 3. Let $\phi : X \to X'$ be a homomorphism of two complex tori of the same dimension. Then ϕ is surjective if and only if the kernel of ϕ is finite.

A homomorphism satisfying the equivalent condition of Exercise 3 is called an isogeny of degree $= \#$ kernel.

§ 1.2 Hermitian Algebra

Let V be a complex vector space. Recall that a Hermitian form H is a pairing $H : V \times V \to \mathbb{C}$ such that $H(z, w)$ is complex linear in z and $H(z, w) = \overline{H(w, z)}$. It follows that $H(z, w)$ is anti-complex linear in w and $H(z, z)$ is a real-valued quadratic form on V.

Let $E(z, w) = \mathrm{Im}\, H(z, w)$ be the imaginary part of a Hermitian form H. Then

a) E is a real skew-symmetric form on V as $\mathrm{Im}\, H(z, z) = 0$ and

b) $E(iz, iw) = E(z, w)$ as $H(iz, iw) = i(-i)\, H(z, w) = H(z, w)$ we may recover H from E.

Proposition 1.2. *Given a form E satisfying a) and b) there is a unique Hermitian form H with imaginary part E.*

Proof. We check the uniqueness first. $H(z, w) = \mathrm{Re}\, H(z, w) + i\, \mathrm{Im}\, H(z, w) = \mathrm{Im}\, H(iz, w) + i\, \mathrm{Im}\, H(z, w)$. Thus

$$(*)\qquad\qquad H(z, w) = E(iz, w) + iE(z, w) \,.$$

This shows uniqueness. Conversely given E define H by the formula. Then we need to check that $H(iz, w) = iH(z, w)$ and $\overline{H(w, z)} = H(z, w)$. These two equations follow easily from the properties a) and b). $\qquad\qquad\square$

We have a relationship between the properties of these forms H and E. Recall that $\operatorname{Ker} H = \{z \in V | H(z, w) = 0 \text{ for all } w \text{ in } V\}$ and $\operatorname{Ker} E = \{z \in V | E(z, w) = 0 \text{ for all } w \text{ in } V\}$.

Lemma 1.3. $\operatorname{Ker} H = \operatorname{Ker} E$.

Proof. As $E = \operatorname{Im} H$, $\operatorname{Ker} E \supset \operatorname{Ker} H$. Conversely by b) $\operatorname{Ker} E$ is invariant under multiplication by i. Thus the formula $(*)$ implies that $\operatorname{Ker} E \supseteq \operatorname{Ker} H$.
$\qquad\square$

Corollary 1.4. *H is non-degenerate if and only if E is non-degenerate.*

Exercise 1. Let $H(z, z) = \sum_{1 \leq j \leq g} z_j \bar{z}_j$ be the standard. Hermitian matrix on $\mathbb{C}^g$. Write the associated Hermitian form $H(z, w)$ and its imaginary part $E(z, w)$.

Exercise 2. In the situation of Lemma 1.3 assume that there is a lattice L in V such that E is integral on $L \times L$. Then $\operatorname{Ker}(H) \cap L$ is a lattice in $\operatorname{Ker} H$ and the image L' of L in $V / \operatorname{Ker} H$ is a lattice. Furthermore H is induced from a non-denegerate Hermitian H' on $V / \operatorname{Ker} H$ which is integral valued on $L' \times L'$.

§ 1.3. The Invertible Sheaves on a Complex Torus

There is a general method for constructing a holomorphic line bundle L'' on the complex torus V/L by taking the quotient of a line bundle on V. Let L' be a line bundle over V. Assume that we have a group action $\alpha : L \times L' \to L'$ such that for any l in L and v in V $\alpha(l, -)$ induces a linear isomorphism from the fiber of L' over v to that over $l + v$. Then the quotient $L'' = L'/L$ is a line bundle over V/L.

In our situation we will consider the case where L' is trivial. Fix an isomorphism $B : \mathbb{C} \times V \to L'$ of bundles over V. Then $\tau = B(1, v)$ is an (arbitrary) nowhere vanishing holomorphic section of L'. Then for any l in L

$$\alpha(l, \tau(v)) = A_l(v) \cdot \tau(l + v)$$

where A_l is a nowhere zero $\mathbb{C}$-analytic function of v in V. The condition that α is a group action is that $\alpha(l_1, \alpha(l_2, \tau(v))) = \alpha(l_1 + l_2, \tau(v))$ or

$$A_{l_1}(v + l_2)A_{l_2}(v)\tau(l_1 + l_2 + v) = \alpha(l_1, A_{l_2}(v)\tau(l_2 + v)) = A_{l_1 + l_2}(v)\tau(l_1 + l_2, v) \, .$$

Therefore we have the cocycle condition

c) $A_{l_1}(v + l_2)A_{l_2}(v) = A_{l_1 + l_2}(v)$ for all l_1, l_2 in L and v in V.

In the prehomology mathematics the multipliers $A_l(v)$ satisfying c) are called factors of automorphy. We are not really interested in line bundles but in invertible sheaves on V/L. For this purpose we will explicitly construct the sheaf $\mathscr{L}$ corresponding to a given factor of automorphy A.

A section of $\mathscr{L}$ over an open subset U of V/L is a holomorphic function $f(v)$ on $\pi^{-1}(U)$ such that $f(l+v) = A_l(v)\,f(l)$ for all l in L and v in $\pi^{-1}(U)$ where $\pi : V \to V/L$ denotes the projection. Clearly $\mathscr{L}$ is a local trivial (i.e., invertible) sheaf on V/L.

There is some simple factors of automorphy that define all invertible sheaves on a complex torus V/L. Appel-Humbert A.-H. data are a pair (α, H) where H is a Hermitian form on V such that its imaginary part E is integral on $L \times L$ and α is a mapping from L to the unitary group $U(1) = \{x \in \mathbb{C}^* \mid |x| = 1\}$ such that

$$\alpha(l_1 + l_2) = \alpha(l_1)\,\alpha(l_2)(-1)^{E(l_1, l_2)} \quad \text{for all } l_1 \text{ and } l_2 \text{ in } L \ .$$

If (α, H) are A.-H. data then one may easily check that $A_l(v) = \alpha(l)e^{\pi H(v,l)+\frac{\pi}{2}H(l,l)}$ are factors of automorphy. Let $\mathscr{L}(\alpha, H)$ be the invertible sheaf on V/L defined by these factors of automorphy.

The significance of this definition is given in the following.

Theorem 1.5. *Any invertible sheaf on V/L is isomorphic to a sheaf of the form $\mathscr{L}(\alpha, H)$ for a uniquely determined Appel-Humbert data (α, H).*

We will prove this theorem in Chapter 3. In this chapter we will explain the significance of this theorem.

Exercise 1. Let $L = Z + Z\tau$ in $C = V$ where τ is a complex number with $\operatorname{Im}\tau > 1$. Determine explicitly all A.-H data for V/L. What is the degree of $L(\alpha, H)$?

Exercise 2. In general show that

$$L(\alpha_1, H) \otimes L(\alpha_2 H_2) \cong L(\alpha_1\alpha_2, H_1 + H_2) \ .$$

Exercise 3. As above, if we change the basic section $\tau(v)$ of L' to $\tau(v)\,B(v)$ where $B(v)$ is as nowhere zero holomorphic function on V. Then the factor of automorphy which describes the same bundle L'' would change to the cohomologous cocycle $A_l(v)\,B(l+v)\,B(l)^{-1}$.

§ 1.4 The Structure of Pic(V/L)

The Picard group Pic(V/L) of a complex torus V/L is the group of isomorphism classes of invertible sheaves on V/L with tensor product as group law. By the last section we know that Pic(V/L) = {A.-H. data for L in V}.

Lemma 1.6. *Given an Hermitian form H on V such that the imaginary part E is integral on $L \times L$ we have exactly $2^{2 \dim_{\mathbf{C}} V} \alpha : L \to \{\pm 1\}$ such that (α, H) are A.-H. data.*

Proof. Let $l_1, \ldots, l_{2g}$ be a basis for the lattice L where $g = \dim V$. Given signs $\alpha(l_i)$ in $\{\pm 1\}$ for each i, let $\alpha(\sum n_i l_i) = \prod \alpha(l_i)^{n_i} (-1)^{\sum_{i<j} n_i n_j E(l_i, l_i)}$. Then one simply checks to (α, H) are exactly the required A.-H. data. $\square$

Let $\mathrm{Pic}^0 = \mathrm{Hom}_{\mathbf{Z}}(L, U(1))$. Then we may regard Pic^0 as a subgroup of Pic(V/L). Here is a homomorphism $\alpha : L \to U(1)$ correspondence to the A.-H. data $(\alpha, 0)$. Thus by Lemma 1.6 we have an exact sequence

$$1 \to \mathrm{Pic}^0 \to \mathrm{Pic}(V/L) \to \{H \text{ such that } \mathrm{Im}\, H(L \times L) \subset \mathbf{Z}\} \to 0 \,.$$

(By construction Pic^0 corresponds to the unitary flat invertible sheaves on V/L.)

Clearly Pic^0 is a real torus. We will next define a complex structure on Pic^0. As a real torus we have the exact sequence

$$0 \to \mathrm{Hom}_{\mathbf{Z}}(L, \mathbf{Z}) \to \mathrm{Hom}_{\mathbf{Z}}(L, \mathbb{R}) \xrightarrow{e^{2\pi i -}} \mathrm{Hom}_{\mathbf{Z}}(L, U(1)) \to 0 \,.$$

Thus we want to put a complex structure on the real vector space $\mathrm{Hom}_{\mathbf{Z}}(L, R)$; i.e. define $i\lambda$ for any homomorphism $\lambda : L \to \mathbb{R}$. Extend λ to a real linear functional $\bar{\lambda}$ on V. We may do this uniquely because L is a lattice in V. Define $i\lambda(l) \equiv -\bar{\lambda}(il)$. Thus $\mathrm{Pic}^0 = \mathrm{Hom}_{\mathbb{R}}(V, R)/\mathrm{Hom}_{\mathbf{Z}}(L, \mathbf{Z})$ is a complex torus dual to V/L. We will need to test the appropriateness of this complex structure of Pic^0.

If $\mu : L \to \mathbf{C}^*$ is a homomorphism, then $\mu(l)$ are constant factors of automorphy for a flat invertible sheaf $M(\mu)$ on V/L. We will check that $M(\mu)$ corresponds to a point $m(\mu)$ of Pic^0. Thus we will have a natural mapping $m : \mathrm{Hom}_{\mathbf{Z}}(L, \mathbf{C}^*) \to \mathrm{Pic}^0$. We intend to check.

Lemma 1.7. *m is a complex analytic homomorphism for the obvious complex group structure on $\mathrm{Hom}_{\mathbf{Z}}(L, \mathbf{C}^*)$.*

Proof. Write $\mu(l) = e^{2\pi i k(l)}$ where $k : L \to \mathbf{C}$ is a homomorphism. Let $\bar{k} : V \to \mathbf{C}$ be the real expansion of k. We have a complex linear mapping $p(\bar{k}) = p :$

$V \to \mathbb{C}$ given by $p(v) = \operatorname{Im} \bar{k}(iv) + i \operatorname{Im} \bar{k}(v)$. Thus $R(\bar{k}) = \bar{k} - p$ is real valued. Now $\mu(l)e^{-2\pi i p(v+l)}(e^{-2\pi i p(v)})^{-1} = e^{2\pi i(\bar{k}-p)(l)} = e^{2\pi i R(\bar{k})(l)}$ is cohomologous to $\mu(l)$ and has values in $U(1)$. Thus $m(\mu)(l) = e^{2\pi i R(\bar{k})(l)}$. We want to prove $R : \operatorname{Hom}_{\mathbb{R}}(V, \mathbb{C}) \to \operatorname{Hom}_{\mathbb{R}}(V, \mathbb{R})$ is a complex linear mapping. Clearly it is real linear. For the rest we want $R(i\bar{k})(v) = -R(\bar{k})(iv)$ but this is readily verified as $R(\bar{k})(v) = \operatorname{Re} \bar{k}(v) - \operatorname{Im} \bar{k}(iv)$. $\qquad\square$

There is an interesting invertible sheaf on the product complex torus $V/L \times \operatorname{Pic}^0(V/L)$ which is called the Poincaré sheaf $\mathscr{P}$. It has the nice property that $\mathscr{P}|_{V/L \times \{(\alpha,0)\}}$ is isomorphic to the sheaf $\mathscr{L}(\alpha, 0)$ on V/L. Consider the Hermitian form $H((u, u'), (v, v')) = -u'(iv) - v'(iu) + i(u'(v) - v'(u))$ where u and v are in V and u' and v' are in $\operatorname{Hom}_{\mathbb{R}}(V, \mathbb{R}) = V\hat{\ }$. The lattice $L\hat{\ }$ in $V\hat{\ }$ which gives Pic^0 is $\{\lambda \in V\hat{\ } | \lambda(L) \subseteq \mathbb{Z}\}$. The imaginary part E of H is $+u'(v) - v'(u)$ and, hence, is integral on $L \times L\hat{\ }$. Let $B : L\hat{\ } \to \mathscr{U}(1)$ be $B(l, l') = (-1)^{l'(l)}$. It is easy to check that (B, H) is A.-H. data for a sheaf $\mathscr{P}$ on $V/L \times \operatorname{Pic}^0(V/L)$.

Exercise 1. Check that the Poincaré sheaf has the above nice property.

Exercise 2. Let $(\alpha(\epsilon), H(\epsilon))$ be a continuous family of A.-H. data on V/L. Show that it has the form $(\beta(t)\gamma, k)$ where $\beta(t)$ is a continuous function with values in Pic^0 and (γ, k) is a constant A.-H. data.

Exercise 3. Let (α, H) be A.-H. data. Then $\alpha(0) = 1$ and $\alpha(l) = \alpha(-l)^{-1}$.

Exercise 4. Show that the (α, H)'s of Lemma 1.6 are exactly those A.-H. data such that $(-1)^*(\mathscr{L}(\alpha, H)) \simeq \mathscr{L}(\alpha, H)$ where $-1 : V/L \to V/L$ is the inverse morphism. (These are called the symmetric sheaves as V/L.)

Exercise 5. Show that the double dual Pic^0 of Pic^0 of V/L is canonically isomorphic to V/L.

Exercise 6. Show that $\dim \operatorname{Pic}^0 = \dim V/L$.

Exercise 7. Let $f : V/L \to V'/L'$ be an isogeny of complex tori. Then we have a natural isogeny

$$f\hat{\ } : \operatorname{Pic}^0(V'/L') \to \operatorname{Pic}^0(V/L)$$

and $\operatorname{Ker}(f\hat{\ })$ is canonically the character group $\operatorname{Hom}(\operatorname{Ker}(f), U(1))$ of $\operatorname{Ker}(f)$. Conclude that

$$\deg f\hat{\ } = \deg f \ .$$

§ 1.5 Translating Invertible Sheaves

In this section we will consider how invertible sheaves change under the action of the group theoretic operations in a complex torus V/L. The idea is to take advantage of the fact that the exponents of the factors of automorphy are linear in v and quadratic in l. This is a much simpler situation than one might imagine without knowledge of the Appel-Humbert Theorem 1.5.

Let x be a point of $X = V/L$. Then $T_x : X \to X$ denotes translation by x; i.e. $T_x(y) = x + y$. Let $\mathscr{L}$ be an invertible sheaf on X. Consider the sheaf $T_x^* \mathscr{L} \otimes \mathscr{L}^{\otimes -1}$ on X. This is a continuous function of x and is trivial when $x = 0$. Thus one expects that there is a point $\varphi_{\mathscr{L}}(x)$ of $\mathrm{Pic}^0(X)$ corresponding to $T_x^* \mathscr{L} \otimes \mathscr{L}^{\otimes -1}$.

Proposition 1.8. $\varphi_{\mathscr{L}} : X \to \mathrm{Pic}^0(X)$ *is a homomorphism of complex tori.*

Proof. We need to find a complex linear mapping $P : V \to \mathrm{Hom}_{\mathbb{R}}(V, \mathbb{R})$ which induces $\varphi_{\mathscr{L}}$. Let $\mathscr{L} = \mathscr{L}(\alpha, H)$ for some A.-H. data (α, H) for V/L. The factors of automorphy $B_l(v)$ describing $T_x^* \mathscr{L} \otimes \mathscr{L}^{\otimes -1}$ are

$$\alpha(l)e^{\pi H(\tilde{x}+v,l)+\pi/2\,H(l,l)} \left(\alpha(l)e^{\pi H(v,l)+\pi/2\,H(l,l)}\right)^{-1} = e^{\pi H(\tilde{x},l)}$$

where $\tilde{x}$ in V is a lifting x. Thus $B_l(v)$ are constant in v and $\varphi_{\mathscr{L}}(x)$ is the point in Pic^0 corresponding to them. In other words we need to compute $\varphi_{\mathscr{L}}(x) = m(B_l)$ by the method of Lemma 1.7.

Now $e^{\pi H(\tilde{x},l)} = B_l$ is $e^{2\pi i(H(\tilde{x},l)/2i)}$. Set $\bar{k}(v) = \frac{1}{2i}H(\tilde{x},l)$. Then $\varphi_{\mathscr{L}}(x)$ is computed by $R(\bar{k})$ which is

$$R(\bar{k})(v) = \mathrm{Re}\left(\frac{1}{2i}H(\tilde{x},v)\right) - \mathrm{Im}\left(\frac{1}{2i}H(\tilde{x},iv)\right)$$

$$= \frac{1}{2}\,\mathrm{Im}\,H(\tilde{x},v) + \frac{1}{2}\,\mathrm{Im}\,H(\tilde{x},v) = E(\tilde{x},v)\,.$$

So $R(\bar{k})$ is a $\mathbb{R}$-linear function of $\tilde{x}$. To check that it is a complex function we need to verify that $E(i\tilde{x},v) = -E(\tilde{x},iv)$ but this follows because $E(i\tilde{x},iv) = E(\tilde{x},v)$. $\qquad\qquad\square$

The proof shows that $\varphi_{\mathscr{L}}$ is exactly the same as E or rather H. In particular we get

Corollary 1.9. *a)* $\varphi_{\mathscr{L}} = 0$ *if and only if* $\mathscr{L}$ *is in* Pic^0, *b)* $\varphi_{\mathscr{L}}$ *is a homomorphism* $\mathrm{Pic}(X) \to \mathrm{Hom}(X, \mathrm{Pic}^0(X))$, *c)* H *is nondegenerate if and only if* $\varphi_{\mathscr{L}}$ *is an isogeny.*

The kernel of $\varphi_{\mathscr{L}}$ is denoted by $K(\mathscr{L})$ and is an important invariant of $\mathscr{L}$. By the above we see that $K(\mathscr{L})$ corresponds to the subgroup of $\tilde{x}$ in V such that $E(\tilde{x}, l)$ is integral for all l in L.

Lemma 1.10. *If H is non-degenerate, $K(\mathscr{L})$ is a finite group of order $\det_L E$.*

Proof. By the above we have $K(\mathscr{L}) \approx \{\tilde{x} \in V | E(\tilde{x}, l) \in \mathbb{Z}\}/L$. Thus the result follows from standard linear algebra. $\square$

Another result along the general lines of discussion of this section is the theorem of the cube.

Theorem 1.11. *If $\mathscr{L}$ is an invertible sheaf on a complex torus X,*
   a) the sheaf $(\pi_1 + \pi_2 + \pi_3)^*\mathscr{L} \otimes (\pi_1 + \pi_2)^*\mathscr{L}^{\otimes -1} \otimes (\pi_1 + \pi_3)^*\mathscr{L}^{\otimes -1} \otimes (\pi_2 + \pi_3)^*\mathscr{L}^{\otimes -1} \otimes \pi_1^*\mathscr{L} \otimes \pi_2^*\mathscr{L} \otimes \pi_3^*\mathscr{L}$ on $X \times X \times X$ is trivial.
   b) For any integer m we have an isomorphism

$$m^*\mathscr{L} \approx \mathscr{L}^{\otimes m(m+1)/2} \otimes (-1)^*\mathscr{L}^{\otimes m(m-1)/2} .$$

Proof. Let (α, H) be A.-H. data for $\mathscr{L}$. Then A.-H. data for the sheaf in part a) are $\big(\alpha(l_1 + l_2 + l_3)\alpha^{-1}(l_1 + l_2)\alpha^{-1}(l_1 + l_3)\alpha^{-1}(l_2 + l_3)\alpha(l_1)\alpha(l_2)\alpha(l_3), H(z_1 + z_2 + z_3, w_1 + w_2 + w_3) - H(z_1 + z_2, w_1 + w_2) - H(z_1 + z_3, w_1 + w_3) - H(z_2 + z_3, w_2 + w_3) + H(z_1, w_1) + H(z_2, w_2) + H(z_3, w_3)\big)$. As H is bilinear the H-part of this data is 0. Thus the α-part is a homomorphism $L \times L \times L \to U(1)$, whose value is clearly one when restricted to the component subgroups. Therefore the α-part is 1 and, hence, a) is true. The part b) can be formally deduced from a) or more directly by reproducing the above type of argument. I leave this detail to the reader. $\square$

Exercise 1. If (α, H) are A.-H. data on a complex torus $X = V/L$ and H is nondegenerate, prove that for some point x of X, $T_x^*\mathscr{L}$ is a symmetric sheaf.

Exercise 2. Let $f : X \to Y$ be a homomorphism between two complex tori and $\mathscr{L}$ is an invertible sheaf on Y. Then we have a commutative diagram

$$
\begin{array}{ccc}
X & \xrightarrow{f} & Y \\
{\scriptstyle \phi_{f^*\mathscr{L}}}\downarrow & & \downarrow{\scriptstyle \phi_{\mathscr{L}}} \\
\mathrm{Pic}^0(X) & \xleftarrow{\hat{f}} & \mathrm{Pic}^0(Y) .
\end{array}
$$

Chapter 2. The Existence
of Sections of Sheaves

§ 2.1 The Sections of Invertible Sheaves (Part I)

Let (α, H) be A.-H. data for a complex torus V/L. Our objective is

Theorem 2.1. *The space* $\Gamma(V/L, \mathscr{L}(\alpha, H))$ *of holomorphic sections of* $\mathscr{L}(\alpha, H)$ *is non-zero if and only if both*
   a) H is positive semi-definite and
   b) α is identically one on $L \cap \operatorname{Ker} H$.

The proof will be given in two sections. Here we will reduce to the most interesting case where H is positive definite.

Step 1. $\Gamma(V/L, \mathscr{L}(\alpha, H)) = 0$ if there exists v in V such that $H(v,v) < 0$. The best idea here is to appeal implicitly to the curvative property of the natural Hermitian metric in $\mathscr{L}(\alpha, H)$. This metric is interesting for other reasons. To define it let $f(v) = e^{-\pi H(v,v)}$. Then f is positive $\mathbb{R}$-analytic function on V and it satisfies the functional equation

$$f(v + l) = e^{-2\pi \operatorname{Re} H(v,l) - \pi H(l,l)} f(v)$$

for all l and v in V. Let $\varphi(v)$ be a local section of $\mathscr{L}(\alpha, H)$ defined near a point of V/L. Then $\|\varphi\|_x^2 \equiv f(w)|\varphi(w)|^2$ for any w in x. This is well-defined because $f(w + l)|\varphi(w + l)|^2 = f(w)|\varphi(w)|^2$ for l in L by the functional equation for f and φ.

Let φ be a global section of $\mathscr{L}(\alpha, H)$. Then as V/L is compact we have a bound M such that $\|\varphi\| \leq M$ for all x in V/L. Therefore for all complex numbers λ, $k(\lambda) = \varphi(\lambda \cdot v)$ is an entire holomorphic function of λ such that $|k(\lambda)| \leq M \cdot e^{-\pi \lambda^2 C}$ where $0 < C = -H(v,v)$. Therefore $k(\lambda) = 0$. In particular $\varphi(v)$ is zero if v is in the non-empty open set of v where $H(v,v) < 0$. There $\varphi = 0$. This proves Step 1.

Step 2. $\Gamma(V/L, \mathscr{L}(\alpha, H)) = 0$ if α is not one on $L \cap \operatorname{Ker} H$.

Let $\varphi(v)$ be a section of $\mathscr{L}(\alpha, H)$. Then for any point v of V and l in $\mathrm{Ker}\, H \cap L$ then $\varphi(v+l) = \alpha(l)\varphi(v)$ where $|\alpha(l)| = 1$. As $\mathrm{Ker}(H) \cap L$ is a lattice in $\mathrm{Ker}(H)$, $\varphi(v)$ is bounded on any $\mathrm{Ker}(H)$-coset in V. Hence $\varphi(v)$ is constant of $\mathrm{Ker}(H)$-cosets. If $\alpha(l) \neq 1$ for all l in $\mathrm{Ker}(H) \cap L$, $\varphi(v) = 0$ for all v. This proves Step 2.

Furthermore if $\alpha(l) = 1$ for all l in $\mathrm{Ker}\, H \cap L$, φ is induced by a function φ' on $V' \equiv V/\mathrm{Ker}(H)$ which satisfies the functional equation to be a section of $\mathscr{L}(\alpha', H')$ where H' is the Hermitian form on V' which induced H and $\alpha' : L' =$ Image of L in $V' \to U(1)$ induces α. (Here one must check that α is constant on $\mathrm{Ker}(H) \cap L$ cosets in L but this is trivial.) As H' is definite by construction and we have shown that $\Gamma(V/L, \mathscr{L}(\alpha, H)) \cong \Gamma(V'/L', \mathscr{L}(\alpha', H'))$ we are reduced to verifying.

Step 3. $\Gamma(V/L, \mathscr{L}(\alpha, H)) \neq 0$ if H is positive definite.

This step is postponed because we want to give an explicit construction of all these sections and will prepare the notation for this in the next section.

Exercise 1. Compute the curvature $\frac{1}{2\pi i}\partial\bar{\partial}\log f$ of our metric on $\mathscr{L}(\alpha, H)$. Show that it is an invariant differential form. Furthermore this is the unique metric with this properly such that $\|1\|_0 = 1$.

Exercise 2. Let X be a non-zero complex torus. Show that the Poincaré sheaf on $X \times \mathrm{Pic}^0$ has no non-zero sections.

§ 2.2 The Sections of Invertible Sheaves (Part II)

Let (α, H) be A.-H. data for a complex torus V/L where $\dim_{\mathbb{C}} V = g$. We will assume that H is positive definite. As usual $E = \mathrm{Im}\, H$. We intend to write the sections of $\mathscr{L}(\alpha, H)$ using theta series. There is some necessary notation.

Let A be a subgroup of the lattice L such that $A = L \cap \mathbb{R} \cdot A$, E is zero on $A \times A$ and rank $A = g$. The existence of such subgroups is easy. If A' satisfies the last two conditions, then $A = L \cap \mathbb{R} \cdot A'$ will work. To find A' choose linearly independent $v_1, \ldots, v_g$ in L inductively such that $E(v_k, v_j) = 0$ for all $j < k$. The set $A' = \mathbb{Z}v_1 + \ldots + \mathbb{Z}v_g$. Thus A is a direct summand of L. Let $W = \mathbb{R} \cdot A \approx \mathbb{R} \otimes_{\mathbb{Z}} A$. Then E is zero on $W \times W$.

The first point is that the canonical mapping $A \otimes_{\mathbb{Z}} \mathbb{C} \to V$ is an isomorphism because $W \cap iW$ is zero as the intersection is a complex subspace on which E (and hence the positive definite form H) is zero.

Because E is zero on $A \times A$ the mapping $\alpha : A \to U(1)$ is a homomorphism. Hence there is a complex linear transformation λ on V such that λ is real on W and $\alpha(a) = e^{2\pi i \lambda(a)}$ for all a in A. Furthermore as H is real on $W \times W$ there is a unique complex symmetric bilinear form S on $V \times V$ such that $H(w_1, w_2) = S(w_1, w_2)$ for all w_1 and w_2 in $W \times W$. By complex linearity $S = H$ on $V \times W$. Also if w is in W and v in V we have $\frac{1}{2i}(H - S)(w, v) = E(w, v)$ because the left side is $\frac{1}{2i}(\overline{H} - S)(v, w) = \frac{1}{2i}(\overline{H} - H) = -\operatorname{Im} H(v, w)$ which is the right side.

For v in V let $\hat{v} : V \to \mathbb{C}$ be the complex linear mapping such that $\hat{v}(w) = E(w, v)$ for all w in W. Thus $\frac{1}{2i}(H - S)(w, v) = \hat{v}(w)$ for all w in W and hence by complex linearity for all w in V. Finally, let $A^{\widehat{\ }} = \operatorname{Hom}_{\mathbb{Z}}(A, \mathbb{Z})$ which will be identified with a subgroup of $\operatorname{Hom}_{\mathbb{C}}(V, \mathbb{C})$.

Let $f(v)$ be a global section of $\mathscr{L}(\alpha, H)$; i.e., f is a holomorphic function on V such that $f(v + l) = \alpha(l)e^{\pi H(v,l) + \frac{\pi}{2}H(l,l)}f(v)$ for all l in L. Consider $g(v) = e^{-\frac{\pi}{2}S(v,v) - 2\pi i \lambda(v)}f(v)$. Then g is still holomorphic but now

$$g(v + l) = \alpha(l)e^{-2\pi i \lambda(l)}e^{\pi(H-S)(v,l) + \frac{\pi}{2}(H-S)(l,l)}g(v)$$

$$= \alpha(l)e^{-2\pi i \lambda(l) + \pi i \hat{l}(l)}e^{2\pi i \hat{l}(v)}g(v) \ .$$

Thus by construction g is periodic with respect to A. Hence we may expand it in a Fourier series $g(v) = \sum_{\chi \in A^{\widehat{\ }}} c_{\chi}e^{2\pi i \chi(v)}$ where the c_{χ} are complex constants. Our functional equation gives

$$\sum_{\chi \in A^{\widehat{\ }}} \left(x_{\chi}e^{2\pi i \chi(l)} \right)e^{2\pi i \chi(v)} = \sum_{\chi \in A^{\widehat{\ }}} \left(c_{\chi}\alpha(l)e^{-2\pi i \lambda(l) + \pi i \hat{l}(l)} \right)e^{2\pi i (\chi + \hat{l})(v)} \ .$$

Setting coefficients equal we get the equations

$$(*) \qquad c_{\chi} = \alpha(l)e^{-2\pi i (\lambda(l) + \chi(l)) + \pi i \hat{l}(l)}C_{\chi - \hat{l}} \text{ for all } l \text{ in } L \text{ and } \chi \text{ in } A^{\widehat{\ }} \ .$$

Conversely once we show that the Fourier series of such c_{χ} converges reasonably we may reverse the process to find all possible sections f.

Let $T(\alpha, H, \lambda, U)$ be the space of all complex valued functions c_{χ} of χ in $U^{\widehat{\ }}$ such that $(*)$ holds. The main result is

Theorem 2.2. *For any c in $T(\alpha, H, \lambda, U)$ the series*

$$\psi_{U,\lambda}(c) = e^{\frac{\pi}{2}S(v,v) + 2\pi i \lambda(v)}\left(\sum_{\chi \in U^{\widehat{\ }}} c_{\chi}e^{2\pi i \chi(v)} \right)$$

converges uniformly on compact subsets to a global section of $\mathscr{L}(\alpha, H)$ and $\psi_{U,\lambda} : T(\alpha, H, \lambda, U) \to \Gamma(V/L, \mathscr{L}(\alpha, H))$ is an isomorphism.

Proof. (Actually the convergence is very fast but the notation is confusing). Let $\|\chi\|$ be a norm on $\operatorname{Hom}_{\mathbb{C}}(V, \mathbb{C})$. We claim that

$$(\dagger) \qquad |c_\chi| \le e^{-C\|\chi\|^2} \quad \text{for all } \chi \text{ in } A^\wedge \text{ for some positive constant } C.$$

First we will show that this implies convergence. Assume that v lies in a given compact subset of V. Then $|\psi_{U,\lambda}(c)| \le E \sum_{\chi \in U^\wedge} |c_\chi| e^{D\|x\|}$ for some constants E and D. Thus the series is majorized by $C \sum_{\chi \in U^\wedge} e^{-C\|\chi\|^2 + D\|x\|}$ whose ratios compare with $C \sum_\chi e^{-C\|\chi\|^2}$. This last series converges similarly to $\sum_{(n_i) \in \mathbf{Z}^n} e^{-\sum n_i^2} = \prod_i (\sum_{n_i \in \mathbf{Z}} e^{-n_i^2})$. Anyway the convergence is very fast.

We will take time proving $(\dagger)$ because we want to develop the structure of $T(\alpha, H, \lambda, U)$ so as to compute its dimension. As A is a direct summand of L we can find a complementary subgroup B with $L = A \oplus B$. The first remark is that in the conditions $(*)$ we need only check them with l in B. This follows because the multipliers $\alpha(a)e^{-2\pi i(\lambda(a)+\chi(a))+\pi i a^\wedge a}$ are 1 by construction when a is in A. The conditions $(*)$ mean that c_χ is determined by the arbitrary constants c_{χ_*} where χ_* are a complete set of cosets representatives of B in $A^\wedge$. Explicitly

$$(\dagger\dagger) \qquad c_{\chi_* + \hat{b}} = \alpha(b')e^{-2\pi i(\lambda(b)+\chi(b))-\pi i b^\wedge(b)} c_{\chi_*}$$

for all b in B and all χ_*.

Next we want to see that the number of χ_* (i.e. $\operatorname{Cok}\{^\wedge : B \to A^\wedge\}$) is finite. Write E as a matrix in terms of the decomposition $L = A \oplus B$. Thus E is given by $\begin{bmatrix} 0 & F \\ -F & G \end{bmatrix}$ where F and G are integral matrices. Here F represents $^\wedge : B \to A^\wedge$. Now $\det E = (\det F)^2$ which is non-zero as E is nondegenerate. Hence $\operatorname{Cok}(^\wedge)$ is a finite group with $(\det E)^{1/2}$ number of elements.

As $|\alpha| = 1$ and $(*)$ is finite then $(\dagger\dagger)$ will imply the estimate $(\dagger)$ if we can show that $\operatorname{Im} b^\wedge(b)$ is a negative definite form on B. This is easy. Just write $b = m + in$ where m and n are in W. If $b \ne 0$ then $n \ne 0$ (otherwise u' would be in $A = W \cap L$). Now $\operatorname{Im} \hat{b}(b) = \operatorname{Im}(\hat{b}(m) + i\hat{b}(n)) = \operatorname{Im}(E(m,b) + iE(n,b)) = E(n,b) = E(n,m) + E(n,in) = E(n,in) = \operatorname{Im} H(n,in) = -H(n,n)$ which is negative as H is positive definite. Thus $(\dagger)$ is true as the series converge. $\square$

We get more information from the proof.

Theorem 2.3. $\dim_{\mathbb{C}} \Gamma(V/L, \mathscr{L}(\alpha, H)) = \sqrt{\det E}$.

As this number is positive we have finished the proof of Theorem 2.1.

Exercise 1. If H is positive definite,

$$\dim \Gamma(V/L, \, \mathscr{L}(\alpha, H)) = \sqrt{\#K(\mathscr{L})} = \sqrt{\deg \phi_{\mathscr{L}(\alpha, H)}} \; .$$

Exercise 2. Let $\mathscr{L}$ be an invertible sheaf of the form $\mathscr{L}(\alpha, H)$ on V/L such that $\mathscr{L}$ has a non-zero section. Show for positive integers m that

$$\dim \Gamma\left(V/L, \, \mathscr{L}^{\otimes m}\right) = \left(\dim \Gamma(V/L, \mathscr{L})\right) m^{(\dim V - \dim \operatorname{Ker} H)} \; .$$

Exercise 3. Let $f : X \to Y$ be an isogeny and $\mathscr{L}$ be an invertible sheaf on Y given by A.-H. data with a positive definite Hermitian form. Then

$$\dim \Gamma(X, \, f^* \mathscr{L}) = \left(\dim \Gamma(Y, \, \mathscr{L})\right)(\deg f) \; .$$

§ 2.3 Abelian Varieties and Divisors

A complex torus V/L is an abelian variety if there exists a positive definite Hermitian form H on V such that its imaginary part is integral on $L \times L$. Such a pair $(V/L, H)$ is called a polarized abelian variety.

Lemma 2.4. *Let X be a complex torus. There is quotient abelian variety Y of X such that the projection $\pi : X \to Y$ is a universal homomorphism from X into an abelian variety.*

Proof. Let $X = V/L$. Let $\mathscr{H}$ be the set of all positive semi-definite Hermitian forms on V with integral imaginary part. Let $K = \bigcap_{H \in \mathscr{H}} \operatorname{Ker} H$. For dimension reasons we may find a finite number $H_1, \dots, H_r$ of elements of $\mathscr{H}$ such that $K = \bigcap_i \operatorname{Ker} H_i$. Let $H = H_1 + \dots + H_r$. Then by construction $\operatorname{Ker} H = K$. Let $Y = V/K//\operatorname{Im} L$. Thus H gives a polarization of Y and hence Y is an abelian quotient. A little thought should give the universal property of Y.   □

Corollary 2.5. *Let $\mathscr{L}$ be an invertible sheaf on X such that $\mathscr{L}$ has a non-zero section. Then $\mathscr{L} = \pi^* \mathscr{M}$ for an invertible sheaf $\mathscr{M}$ on Y and*

$$\pi^* : \Gamma(Y, \mathscr{M}) \to \Gamma(X, \mathscr{L})$$

is an isomorphism.

Proof. Consult Step 2 of the proof of Theorem 2.1.   □

Next we will explain our results in terms of divisors. Recall that a divisor on a complex manifold X is an element of the free abelian group on irreducible divisors which are the irreducible closed subvarieties of X of codimension 1. A divisor D is effective if all its multiplicities are non-negative.

We have a homomorphism $\{$Divisor on $X\} \to \mathrm{Pic}(X)$ which sends D to the sheaf $\mathcal{O}_X(D)$. Two divisors D_1 and D_2 are linearly equivalent (written $D_1 \sim D_2$) if and only if by definition $\mathcal{O}_X(D_1) \approx \mathcal{O}_X(D_2)$ if and only if there exists a non-zero meromorphic function f on X such that the divisor (f) of zeros and poles of f equals $D_1 - D_2$. The complete linear system $|D|$ of a divisor D consists of all effective divisors linearly equivalent to D. If X is compact there is a natural bijection between $|D|$ and the projective space of lines in $\Gamma(X, \mathcal{O}_X(D))$.

Let X be a complex torus and $\pi : X \to Y$ be its abelianization.

Corollary 2.6. *The homomorphism π induces a bijection between*

    a) $\{$divisors on $Y\}$ and $\{$divisors on $X\}$ which respects linear equivalence, *and*

    b) $\{$meromorphic functions on $Y\}$ and $\{$meromorphic functions on $X\}$.

Proof. This is a reformulation of Corollary 2.5 in classical language. I leave it as an exercise for those who want to speak both languages. $\square$

Therefore if you are just interested in the geometry of divisors or the algebra of meromorphic functions on the complex torus you need only work with abelian varieties.

A useful fact about divisors is the theorem of the square which is a special case of Proposition 1.8.

Proposition 2.7. *For any divisor D on a complex torus X and point x and y of X, $(D + x + y) + (D) \sim (D + x) + (D + y)$.*

Exercise 1. Show that any complex torus of dimension one is an abelian variety. In fact there is a canonical choice of a polarization.

Exercise 2. Let $l_1 = (1, 0)$, $l_2 = (0, 1)$, l_3, l_4 be a basis for a lattice $L(l_3, l_4)$ in $\mathbb{C}^2$. For fixed l_3, l_4 let H be a polarization of $\mathbb{C}^2/L(l_3, l_4)$. Vary l_3 and l_4 in $\mathbb{C}^2$ to $\tilde{l}_3$ and $\tilde{l}_4$. Show that there is one non-trivial $\mathbb{C}$-analytic condition on $\tilde{l}_3$, $\tilde{l}_4$ such that there is a polarization $\tilde{H}$ on $\mathbb{C}^2/L(\tilde{l}_3, \tilde{l}_4)$ with imaginary part = the analytic continuation of the imaginary part of H.

§ 2.4 Projective Embeddings of Abelian Varieties

We will begin with two simple consequences of the theorem of the square. Let D be an effective divisor on a complex torus.

Lemma 2.8. *If n is an integer ≥ 2 then the linear system $|nD|$ has no base points.*

Proof. Let y be a given point of X. We want to find a divisor E in $|nD|$ which does not pass through y. Now y is contained in $D + x$ if and only if x is contained in the codimension one subvariety $y - D$. Thus for general choices of $x_1,\ldots,x_{n-1}$, y is contained in $(D + x_1) + (D + x_2)+\ldots+(D + x_{n-1}) + (D - \sum_{1\leq j\leq n-1} x_j) = E$. By the theorem of the square 2.7 E is contained in $|nD|$. $\qquad\qquad\square$

Lemma 2.9. *The linear system $|D|$ contains reduced divisors.*

Proof. Let $D = \sum n_j D_j$ where D_i are the components. If $n_j > 1$ replace $n_j D_j$ by a divisor E_j formed from D_j as in the last lemma. Thus $D \sim \sum E_j = E$. We want to choose the x's for E_j general so that E is reduced. This follows by the following maneuver. Given two non-empty effective divisors F_1 and F_2 then the set of x such that $F_1 + x = F_2$ is contained in the divisor $F_2 - f_1$ where f_1 is any point of F_1. $\qquad\qquad\square$

Let (α, H) be A.-H. for the sheaf $\mathcal{O}_X(D)$ on X. We begin the serious discussion with

Proposition 2.10. *The subset $\{E \in |D| \mid E + x = E,$ for some non-zero x in $X\}$ of $|D|$ is the union of a finite number of proper linear subspaces if H is positive definite.*

Proof. Assume that $E = x + E$ for some E in $|D|$. Thus $T^*_{-x}\mathcal{O}_X(E) \approx \mathcal{O}_X(E)$. Hence x is contained in the finite group $K(\mathcal{O}_X(D))$. In particular x generates a finite subgroup S of X. Let $\pi : X \to X/S = Y$ be the quotient homomorphism. There is a divisor F on Y such that $E = \pi^{-1}F$. We write $X = V/L$ and $Y = V/M$. Let (α, H) be A.-H. data for the sheaf $\mathcal{O}_Y(F)$. As $(\alpha|_L, H)$ is A.-H. data for $\mathcal{O}_X(E) \approx \mathcal{O}_X(D)$, H and $\alpha|_L$ are fixed. Clearly α is determined by the value $\alpha(l')$ where l' is a lifting of x. Furthermore nl' is contained in L for some integer n (= order x) and $\alpha(l')^n = \alpha(nl')$. Therefore there are only a finite number of choices of α. By construction the cone over $\{E = x + E\}$ in $|D|$ is union of the images of all $\Gamma(Y, \mathscr{L}(\alpha, H))$ in $\Gamma(X, \mathscr{L}(\alpha|_L, H))$. The remaining point is that these finitely many subspaces are proper if $x \neq 0$ as $\deg \pi > 1$ (see exercise on §2.2). $\square$

Now we are in a position to prove the classical theorem of Lefschetz.

Theorem 2.11. *If H is positive definite and $n \geq 3$ then the linear system $|nD|$ defines a projective embedding $\varphi : X \to \mathbb{P}^n$.*

Proof. By Lemma 2.8 φ is defined. By Lemma 2.9 and Proposition 2.10 we may assume that the effective divisor D is reduced and $D + z = D$ implies that $z = 0$. We do the case $n = 3$ to simplify the notation. We will first show that φ separates points; i.e. given two distinct points x and y of X we may find a divisor E in $|3D|$ which passes through x but not y. Consider the divisor $E(a, b) = (D + a) + (D + b) + (D - a - b)$ in $|3D|$. We want to prove that if x is in $E(a, b)$ if and only if y is in $E(a, b)$ then $x = y$. Now if x is in $D + a$ then y is in $E(a, b)$ for any b. Thus y is in $D + a$. So by symmetry x is in $D + a \Leftrightarrow y$ in $D + a$, or, rather $x - D = y - D$. Thus $(-x + y) + D = D$. Hence $-x + y = 0$ or $x = y$. Thus φ is injective.

Let r be a non-zero tangent vector in X. We need to find a divisor $E(a, b)$ passing through a fixed point x but which is not tangent to r at x. Take a in $x - D$ and b general then $D + a$ contains x but $(D + b) + (D - a - b)$ does not contain x. Thus we will be done unless $D + a$ is tangent to r at x for all a in $x - D$. By translation this means that D is tangent to r at all its points. We need to see that this is impossible.

Let (α, H) be A.-H. data for $\mathcal{O}_X(D)$ where $X = V/L$. Let $f(v)$ be a section of $\mathcal{L}(\alpha, H)$ with zero divisor D. We have $\frac{d}{dr} f(v) = 0$ when $f(v) = 0$. As D is reduced this means that $\frac{1}{f} \frac{d}{dr} f(v)$ is an entire function say g on V. We intend to prove that $r = 0$. This will be our needed contradiction. Let $A_l(v) = \alpha(l) e^{\pi H(v, l) + \frac{\pi}{2} H(l, l)}$ be the factor of automorphy for (α, H). Set $b_l = \frac{1}{A_l(v)} \frac{d}{dr} A_l(v) = \pi H(r, l)$. From the functional equation of f we have $g(v + l) = g(v) + b_l$ for all l in L and v in V. Thus dg is a holomorphic differential on the compact V/L. Hence $dg = dR$ where R is a complex linear function on V. So $g = R + \text{constant}$ and hence $b_l = R(l)$. Thus $\pi H(r, l) = R(l)$ for all l in L where the first expression is anticomplex linear in l and the second is complex linear. Therefore $\pi H(r, ^*) = 0$ for all *. Hence $r = 0$ as H is non-degenerate. Thus φ is infinitesimally injective. $\square$

We have much control of the existence of abelian functions now.

Corollary 2.12. *Let X be complex torus of dimension g. The following are equivalent:*

a) X *is an abelian variety,*

b) *there are g algebraically independent meromorphic functions on X, and*

c) X *is complex projective variety.*

Proof. The Theorem 2.10 proves that a) implies c). By standard algebraic geometry c) implies b). In fact the transcendence degree of the field of algebraic functions on a projective variety equals its dimension. To prove that b) implies a). Let $\pi : X \to Y$ be the abelianization of X. By Corollary 2.6, X and Y has the same meromorphic functions but by the above we know that the transcendence degree of the function field on $Y = \dim Y$. This equals $\dim X$ if and only if $X = Y$. Thus a) is true if b) is true. $\square$

Exercise 1. Show that Lemma 2.7 and Theorem 2.10 are sharp when X has dimension one.

Chapter 3. The Cohomology of Complex Tori

§ 3.1 The Cohomology of a Real Torus

Let L be a lattice in a real vector space W. Then the quotient W/L is a real torus. Using a basis $l_1,\dots,l_2$ of L we have an isomorphism $\varphi : (\mathbb{R}/\mathbb{Z})^n = \mathbb{R}/\mathbb{Z}^n \xrightarrow{\approx} W/L$ which sends a real n-vector $(\lambda_1,\dots,\lambda_n)$ to the L-coset of $\sum \lambda_j l_j$. Thus a real torus is topologically isomorphic to a product of circles. Consequently the algebraic topology of V/L is obvious.

Explicitly for all subset S of $[1,\dots,n]$ we have an oriented $\#S$-cycle $\sigma(S)$ in W/L which is the image via φ of the subset $\{(\lambda_i) \in [0,1]^n | \lambda_t = 0 \text{ if } t \notin S\}$. Clearly the cycles $\sigma(S)$ are a basis for homology of W/L.

Let ω be a closed differential i-form and σ be an oriented i-cycle. The period $\int_\sigma \omega$ only depends on the homology class of σ and the DeRham class of ω modulo exact forms. Thus we have a pairing

$$\int : H_i(W/L, \mathbb{Z}) \times H^i_{\mathrm{DeRham}}(W/L, \mathbb{C}) \to \mathbb{C} .$$

By DeRham's theorem the pairing identifies $H^i_{\mathrm{DeRham}}(W/L, \mathbb{C})$ with

$$\mathrm{Hom}_{\mathbb{Z}}(H_i(W/L, \mathbb{Z}), \mathbb{C}) \approx H^i(W/L, \mathbb{C}) .$$

The integral cohomology $H^i(W/L, \mathbb{Z})$ is identified with the subset of classes in $H^i_{\mathrm{DeRham}}(W/L, \mathbb{C})$ which have integral periods.

Consider the invariant (under translation) differential i-form ω on V/L. Then ω is determined by its value at zero which is an arbitrary element of $\Lambda^i(\mathrm{Hom}_{\mathbb{R}}(W, \mathbb{C}))$. Using coordinates one can easily check the truth of

Lemma 3.1. *a) Any invariant form is closed and each DeRham class contains a unique invariant form.*

b) We have a natural isomorphism $\Lambda^i(\mathrm{Hom}_{\mathbb{R}}(W, \mathbb{C})) \xrightarrow{\approx} H^i_{\mathrm{DeRham}}(W/L, \mathbb{C})$ such that $H^i(W/L, \mathbb{Z})$ is identified with $\Lambda^i(L^\wedge)$ where

$$L^\wedge = \{\varphi \in \mathrm{Hom}_{\mathbb{R}}(W, \mathbb{C}) | \varphi(L) \subseteq \mathbb{Z}\} .$$

Next we consider the Riemannian geometry of W/L. Assume that we are given an invariant Riemannian metric on W/L or what is the same as the Euclidean metric on W. We have the usual $*$-operator on differential forms. As the metric is invariant $*$ of an invariant form is invariant.

Recall that a form ω is called coclosed if $d^*\omega = 0$. As W/L is compact, a form is harmonic if it is closed and coclosed. Thus the invariant forms are harmonic and conversely!

Lemma 3.2. *On the real torus W/L the harmonic forms are exactly the invariant forms.*

Proof. By standard Riemannian geometry on compact manifolds any DeRahm class contains a unique harmonic form. $\qquad\qquad\square$

Exercise 1. Let ω be a one-form on W/L. Then ω is invariant if and only if $\mu^*\omega = \pi_1^*\omega + \pi_2^*\omega$ where $\mu : W/L \times W/L \to W/L$ in the group law.

§3.2 A Complex Torus as a Kähler Manifold

A Riemannian metric on a complex manifold X is Hermitian if it has the form $H(v,v)$ where H is a Hermitian form on the tangent space of X with respect to its given complex structure. The metric is called Kähler if the 2-form (Kähler form) $\operatorname{Im} H(v,\omega)$ is closed.

Let V/L be a complex torus. Let H be a Hermitian form on the tangent space V. Then H extends uniquely to an invariant Hermitian metric on V/L. As the Kähler form is automatically invariant it is closed. Thus we have a Kähler metric on the compact complex manifold V/L and we may use the properties of the Hodge decomposition.

We need to write a given invariant (harmonic) differential form into its (p,q)-components. This is very easy. As a $\mathbb{C}$-valued $\mathbb{R}$-linear function on V can be written uniquely in the form $l_1 + \bar{l}_2$ where the l's are complex linear we have an isomorphism of $\mathbb{C}$-vector spaces

$$\operatorname{Hom}_{\mathbb{R}}(V, \mathbb{C}) = \operatorname{Hom}_{\mathbb{C}}(V, \mathbb{C}) \oplus \overline{\operatorname{Hom}_{\mathbb{C}}(V, \mathbb{C})} \ .$$

Therefore $\Lambda^j\left(\operatorname{Hom}_{\mathbb{R}}(V, \mathbb{C})\right) = \bigoplus_{p+q=j} \Lambda^p\left(\operatorname{Hom}_{\mathbb{C}}(V, \mathbb{C})\right) \otimes \Lambda^q\left(\overline{\operatorname{Hom}_{\mathbb{C}}(V, \mathbb{C})}\right)$.
In terms of cohomology this is the Hodge decomposition $H^i_{\mathrm{DeRham}}(V/L, \mathbb{C}) = \bigoplus_{i=p+q} H^{p,q}$. As this is independent of H we may forget it. As $H^i(V/L, \mathcal{O}_{V/L}) = H^{0,i}$ Hodge theory gives

Theorem 3.3. $H^i(V/L, \mathcal{O}_{V/L}) \cong \Lambda^i\left(\overline{\operatorname{Hom}_{\mathbf{C}}(V, \mathbf{C})}\right)$ *where we associate the Dolbeault cohomology class to the appropriate invariant $(0, i)$-form.*

We will need simple calculation. Let $H(v, w)$ be a Hermitian form on V. Then $\frac{i}{2}\partial\bar{\partial}H(v, v)$ is an invariant $(1, 1)$-form on V and hence we may regard it as one on V/L. We simply want to compute the element of $\Lambda^2\left(\operatorname{Hom}_{\mathbb{R}}(v, \mathbf{C})\right)$ corresponding to it. The result is

Lemma 3.4. *a)* $\frac{i}{2}\partial\bar{\partial}H(v, v)$ *is the invariant form corresponding to the skew-symmetric form* $\operatorname{Im} H(v, w)$.

b) A real skew form $E(e, w)$ on V has type $(1, 1)$ if and only if $E(iv, iw) = E(v, w)$.

c) All real invariant $(1, 1)$ forms on V/L may be written uniquely in the form $\frac{i}{2}\partial\bar{\partial}H'(v, v)$ where H' is a Hermitian form on V.

Proof. b) is easy. It says that the space of $(1, 1)$-form is the 1-eigenspace for the operator $E(v, w) \to E(iv, iw)$. The form of type $(0, 2)$ or $(2, 0)$ are -1-eigenvector for the operator. Thus the statement is clear. For a) we compute in coordinates $(z_1, \ldots, z_g)$ on V. We may assume that H is diagonal, i.e. $H(z, z) = \sum a_k z_k \bar{z}_k$ where the a_k's are real. The $\operatorname{Im} H(z, w) = \sum a_j(y_j u_j - x_j v_j)$ and $\partial\bar{\partial}H(z, z) = \sum a_j dz_j \wedge d\bar{z}_j = i \sum a_j(dy_j \wedge dx_j - dx_j \wedge dy_j)$. Thus the result is clear. The point c) is a combination of a) and b) by Proposition 1.2. $\square$

§3.3 The Proof of the Appel-Humbert Theorem

Let $X = V/L$ be a complex torus. Let A.-H. be the group of all Appel-Humbert data and $\operatorname{Pic}(X)$ the group of isomorphism classes of invertible sheaves on X. Then we have a homomorphism $\rho : \text{A.-H} \to \operatorname{Pic}(X)$ which sends (α, H) to the isomorphism class of $\mathscr{L}(\alpha, H)$. The Appel-Humbert theorem says that ρ is an isomorphism.

Step 1. ρ is injective.

If (α, H) is in the kernel, $\mathscr{L}(\alpha, H)$ is trivial. Thus $\mathscr{L}(\alpha, H)$ and its inverse $\mathscr{L}(\alpha^{-1}, -H)$ have a non-zero section. Thus by Theorem 2.1 H and $-H$ are both positive semi-definite. Therefore $H = 0$ and $L \cap \operatorname{Ker} H = L$. Hence by the theorem again α is identically one. This completes Step 1.

We first recall some generalities. The Picard group $\operatorname{Pic}(X)$ is isomorphic to $H^1(X, \mathcal{O}_X^*)$ where $\mathcal{O}_X^*$ is the sheaf of holomorphic units. We have the exact sequence of sheaves $0 \to \mathbb{Z} \to \mathcal{O}_X \xrightarrow{\text{exp}} \mathcal{O}_X^* \to 0$ which gives an exact sequence of groups

$$(*) \qquad H^1(X, \mathcal{O}_X) \xrightarrow{\ i\ } H^1(X, \mathcal{O}_X^*) \xrightarrow{\ \delta\ } H^2(X, \mathbb{Z}) \ .$$

If $\mathcal{L}$ is an invertible sheaf $\delta[\mathcal{L}] = c_1(\mathcal{L})$ is the (first) Chern class of $\mathcal{L}$ and it determines $\mathcal{L}$ topologically. As $H^2(X, \mathbb{Z}) \subset H^2(X, \mathbb{C})$ we have computed $c_1(\mathcal{L})$ differentiably as below.

Choose a metric on $\mathcal{L}$. Then $c_1(\mathcal{L})$ is represented by $\frac{1}{2\pi i}$ of the curvature $\imath$ of the metric, which is a 2-form. Explicitly if $\hbar$ is the square length of a local holomorphic section of $\mathcal{L}$ then $\imath = \partial\overline{\partial}\log \hbar$. Therefore a Chern class is an integral $(1,1)$-form.

Step 2. $\operatorname{Im}\delta = \operatorname{Im}\delta\rho$.

We will show that any integral $(1,1)$-form ω is cohomologous to the first Chern class of $\mathcal{L}(\alpha, H)$ for some (α, H) in A.-H. Such a two-form ω is cohomologous to the invariant differential of a skew-symmetric form E on V such that $E(iv, iw) = E(v, w)$ by Lemma 3.4 b). Let H be the Hermitian form on V with imaginary part E. As ω is integral, E is integral on $L \times L$. Thus by Lemma 1.6 there exist α such that (α, H) is in A.-H. This step will be finished by applying.

Lemma 3.5. *If* $\mathcal{L} = \mathcal{L}(\alpha, H)$, *then* $\frac{1}{2\pi i}$ *of the curvature of the canonical metric on* $\mathcal{L}$ *is the invariant* $(1,1)$ *form corresponding to* $E = \operatorname{Im} H$.

Proof. The square length of the Section 1 of $\mathcal{L}$ is $e^{-\pi H(v,v)}$. Thus the Chern class $\delta[\mathcal{L}]$ is $\frac{1}{2\pi i}\partial\overline{\partial}\log\left(e^{-\pi H(v,v)}\right) = \frac{i}{2}\partial\overline{\partial}H(v,v)$ which is essentially E by Lemma 3.4 a). $\qquad\square$

We will be done if we can complete

Step 3. $\operatorname{Ker}\delta \subset \operatorname{Im}\rho$.

By the sequence $(*)\operatorname{Ker}\delta = \operatorname{Image} i$. Let β be an element of $H^1(X, \mathcal{O}_X)$. Then by Theorem 3.3, β is the cohomology class of an invariant $(0,1)$-form ω which we may identify with a anti-complex linear function k on V. The general fact is that $i(\beta)$ is represented by the flat sheaf with multipliers $\exp(k(l))$ for l in L. (This is proven as follows. Note that β is the image of the cohomology class γ in $H^1(X, \mathbb{C})$ represented by k. Then $i(\beta)$ is the image of the element $\exp(\gamma)$ of $H^1(X, \mathbb{C}^*)$.) We have seen in the proof of Lemma 1.7 that any flat sheaf is isomorphic to one of the form $\mathcal{L}(\alpha, 0)$ where $(\alpha, 0)$ is in A.-H. Thus $i(\beta)$ is contained in the image of ρ and we are finished.

Remark. It is a general theorem about compact Kähler manifolds that the image of $H^1(X, \mathcal{O}_X^*)$ in $H^2(X, \mathbb{C})$ is exactly the integral $(1,1)$ classes and the image of $H^1(X, \mathcal{O}_X)$ in $H^1(X, \mathcal{O}_X^*)$ is represented by unitary flat bundles.

§ 3.4 A Vanishing Theorem
for the Cohomology of Invertible Sheaves

Let $X = V/L$ be a complex torus. Let (α, H) be A.-H. data. Let A^0 be the space of all C^∞ sections of $\mathcal{L}(\alpha, H)$. Thus A^0 consists of all C^∞-functions f on V such that $f(v + l) = \alpha_l e^{\pi H(v,l) + \frac{\pi}{2} H(l,l)} f(v)$ for all l in L. We have a inner product $(f, g) = \int_X e^{-\pi H(v,v)} f(v) \overline{g(v)} dv$ where dv is the invariant measure on V.

We will consider some differential operators on A^0. Let $(z_1, \ldots, z_g)$ be coordinates on V. Then we may assume that H is diagonal; i.e., $H(z, w) = \sum_{1 \leq j \leq g} h^j z_j \bar{w}_j$ for some real numbers $h^1, \ldots, h^g$. The first operator is just $\left(\frac{\partial}{\partial \bar{z}_j} \right)$. As the multipliers of $\mathcal{L}(\alpha, H)$ are complex analytic in v, the product formula shows that $\frac{\partial}{\partial \bar{z}_j}$ maps A^0 into A^0. We have adjoint operators.

Lemma 3.6. *a) The operator* $\left(\frac{\partial}{\partial \bar{z}_j} \right)^* = -\frac{\partial}{\partial z_j} + \pi h^j \bar{z}_j$ *maps* A^0 *into* A^0.

b) $\left(\left(\frac{\partial}{\partial \bar{z}_j} \right)^* f, g \right) = \left(f, \left(\frac{\partial}{\partial \bar{z}_j} \right) g \right)$ *and* $\left(\left(\frac{\partial}{\partial \bar{z}_j} \right) f, g \right) = \left(f, \left(\frac{\partial}{\partial \bar{z}_j} \right)^* g \right)$.

c) $\left(\frac{\partial}{\partial \bar{z}_j} \right) \left(\frac{\partial}{\partial \bar{z}_j} \right)^* - \left(\frac{\partial}{\partial \bar{z}_j} \right)^* \left(\frac{\partial}{\partial \bar{z}_j} \right) = \pi h^j$.

Proof. For a) let f be an element of A^0. Then

$$\left(\frac{\partial}{\partial \bar{z}_j} \right)^* f(z + l) = -\frac{\partial}{\partial z_j} f(z + l) + \pi h^j (\bar{z}_j + \bar{l}_j) f(z + l)$$

$$= -\alpha_l e^{\pi H(z,l) + \frac{\pi}{2} H(l,l)} \frac{\partial}{\partial z_j} f(z) - \alpha_l e^{\pi H(z,l) + \frac{\pi}{2} H(l,l)} \pi h^j \bar{l}_j f(z)$$

$$+ \left[\pi h^j \bar{z}_j + \pi h^j \bar{l}_j \right] \alpha_l e^{\pi H(z,l) + \frac{\pi}{2} H(l,l)} f(z)$$

$$= \alpha_l e^{\pi H(z,l) + \frac{\pi}{2} H(l,l)} \left(-\frac{\partial}{\partial z_j} f(z) + \pi h^j \bar{z}_j f(z) \right)$$

$$= \alpha_l e^{\pi H(z,l) + \frac{\pi}{2} H(l,l)} \left(\frac{\partial}{\partial \bar{z}_j} \right)^* f .$$

Thus a) is true.

To prove b) it will be enough by Stokes' theorem to show that

$$\left(\left(\frac{\partial}{\partial \bar{z}_j} \right) f, g \right) - \left(f, \left(\frac{\partial}{\partial \bar{z}_j} \right)^* g \right) = \frac{\partial}{\partial \bar{z}_j} (f, g)$$

and

$$\left(\left(\frac{\partial}{\partial \bar{z}_j}\right)^* f, g\right) - \left(f, \left(\frac{\partial}{\partial \bar{z}_j}\right) g\right) = -\frac{\partial}{\partial \bar{z}_j}(f, g)$$

where the inner product is the pointwise version. Now

$$\frac{\partial}{\partial \bar{z}_j}(f, g) = \frac{\partial}{\partial \bar{z}_j}\left(e^{-\pi H(z,z)} f(z)\overline{g(z)}\right)$$

$$= -\pi h^j \bar{z}_j e^{-\pi H(z,z)} f(z)\overline{g(z)} + e^{-\pi H(z,z)}\left(\frac{\partial}{\partial \bar{z}_j}\right) f(z)\overline{g(z)}$$

$$+ e^{-\pi H(z,z)} f(z)\frac{\overline{\partial}}{\partial \bar{z}_j} g$$

$$= -\left(\left(\frac{\partial}{\partial \bar{z}_j}\right)^* f, g\right) + \left(f, \left(\frac{\partial}{\partial \bar{z}_j}\right) g\right) .$$

This proves the first equation. The second equation is conjugate to the first. This proves b). For c) just apply the operators to a function. $\square$

Next we let $A^* = \bigoplus_{0 \leq n \leq g} A^n$ be the space of all C^∞ $\mathscr{L}(\alpha, H)$-valued anti-holomorphic differentials. Let $I = (i_1 < \ldots < i_n)$ be a subset of $[1, \ldots, g]$. The set of all such subsets is Ind. For I in Ind let $d\bar{z}_I = d\bar{z}_{i_1} \wedge \ldots \wedge d\bar{z}_{i_n}$. Then an element of A^n has the form $\omega = \sum_{\substack{I \subset \text{Ind} \\ \#I=n}} \omega_I d\bar{z}_i$ where ω_I is in A^0.

As with $\left(\frac{\partial}{\partial \bar{z}_j}\right)$ the operator $\overline{\partial}$ on differential forms sends A^n to A^{n+1} and $\overline{\partial}^2 = 0$. We want an expression for the adjoint $(\overline{\partial})^*$ of $\overline{\partial}$. Consider the linear operator defined by $(\overline{\partial})^*(f_I d\bar{z}_I) = \sum_{1 \leq d \leq \#I}(-1)^{d+1}\left(\frac{\partial}{\partial \bar{z}_{i_d}}\right)^* f_I d\bar{z}_{I-\{i_d\}}$.

Lemma 3.7. *a)* $(\overline{\partial}^*)$ *takes* A^n *into* A^{n-1}.

b) $(\overline{\partial})^*$ *is adjoint to* $(\overline{\partial})$.

c) $((\overline{\partial})^*\overline{\partial} + \overline{\partial}(\overline{\partial})^*)(f_I d\bar{z}_I) = (\sum_{1 \leq l \leq g}(\frac{\partial}{\partial \bar{z}_l})^*(\frac{\partial}{\partial \bar{z}_l}) + \pi \sum_{j \in I} h^j)f_I d\bar{z}_I$.

Proof. a) is clear from Lemma 3.6. Also b) follows from the lemma and the explicit formula for $\overline{\partial}$ given by $\overline{\partial}(f_I dz_I) = \sum_{1 \leq k \leq g} \frac{\partial}{\partial \bar{z}_k}(f)_I d\bar{z}_k \wedge d\bar{z}_I$. For c) there is a usual sign cancellation which implies that the right hand expression is a multiple of $d\bar{z}_I$. The multiple is $(\sum_{j \in I}(\frac{\partial}{\partial \bar{z}_j})(\frac{\partial}{\partial \bar{z}_j})^* + \sum_{k \notin I}(\frac{\partial}{\partial \bar{z}_k})^*(\frac{\partial}{\partial \bar{z}_k}))(f)$. By the Lemma 3.6, this is the left hand side. $\square$

The cohomology $H^i(X, \mathscr{L}(\alpha, H))$ on $X = \mathbb{C}^g/L$ which we may consider as a Kähler manifold is isomorphic to Ker $\Delta : A^i \to A^i$ where $\Delta = (\overline{\partial})^*\overline{\partial} + \overline{\partial}(\overline{\partial})^*$ is the Laplacian. Thus by Lemma 3.7 c) as Δ respects the decomposition of forms Ker $\Delta = \bigoplus_{\substack{I \in \text{Ind} \\ \#I=i}} H_I$ where

$$H_I = \operatorname{Ker} \Delta : A^0 d\bar{z}_I \to A^0 d\bar{z}_I$$

$$= \left\{ f \in A^0 \,\Big|\, \sum_{1 \le l \le g} \left(\frac{\partial}{\partial \bar{z}_l}\right)^* \left(\frac{\partial}{\partial \bar{z}_l}\right) f = \pi \sum_{j \in I} h^j f \right\} d\bar{z}_I .$$

Lemma 3.8. *Let* $N = \{1 \le j \le g | h^j < 0\}$ *and* $SN = \{1 \le j \le g | h^j \le 0\}$. $H_I = 0$ *if either* $I \not\subseteq SN$ *or* $I \not\supseteq N$.

Proof. If we apply Serre duality we see that $H_I(\mathscr{L}(\alpha, H))$ is dual to $H_{[1,\ldots,g]-I}$ $(\mathscr{L}(\alpha^{-1}, -H))$. Thus the first alternative for $\mathscr{L}(\alpha^{-1}, -H)$ implies the second alternative for $\mathscr{L}(\alpha, H)$. Thus we need to see that if we have k in I such that $h^k > 0$ then $H_I = 0$. Let $\omega = f d\bar{z}_I$ be an element of $H_I = \operatorname{Ker} D$. Then

$$0 = (\Delta \omega, \omega) = \sum_{1 \le l \le g} \left(\frac{\partial}{\partial \bar{z}_l}\right)^* \left(\frac{\partial}{\partial \bar{z}_l}\right) f + \pi \sum_{j \in I} h^j (f, f)$$

$$= \sum_{1 \le l \le g} \left(\frac{\partial f}{\partial \bar{z}_l}, \frac{\partial f}{\partial \bar{z}_l}\right) + \pi \sum_{j \in I} h^j (f, f) .$$

Thus $\sum_{j \in I} h^j < 0$ if $\omega \ne 0$. Thus we are done if h^k is bigger than $\sum_{j \in I - k} h^j$. We may make a diagonal change of coordinates such that this is true and not change the cohomology upto isomorphism. Thus $H_I = 0$. $\qquad\square$

Corollary 3.8. $H^i(X, \mathscr{L}) = \bigoplus_{\substack{N \subseteq I \subseteq SN \\ \#I = i}} H_I$.

§ 3.5 The Final Determination
of the Cohomology of an Invertible Sheaf

Let (α, H) be A.-H. data for an invertible sheaf $\mathscr{L}$ on a complex torus $X = V/L$. Let $z = \dim \operatorname{Ker} H$. Let $Y = V/\operatorname{Ker}(H) + L = (V/\operatorname{Ker} H)/L'$. Then H is induced from a Hermitian definite form H' on $V/\operatorname{Ker} H$. Let n be the number of negative eigenvalues of H. Let $K^0(\mathscr{L}) = \operatorname{Ker} H/L \cap \operatorname{Ker} H$ as usual. Then we have our objective

Theorem 3.9. *a)* $H^i(X, \mathscr{L}) = 0$ *if* $i < n$ *or* $i > n + z$.
 b) *If* $0 \le i \le z$, $H^{n+i}(X, \mathscr{L}) \approx H^n(X, \mathscr{L}) \otimes H^i\left(K^0(\mathscr{L}), \mathcal{O}_{K^0(\mathscr{L})}\right)$.
 c) *If* $\alpha | L \cap \operatorname{Ker} H \not\equiv 1$, *then* $H^n(X, \mathscr{L}) = 0$ *and, otherwise,* $\dim H^n(X, \mathscr{L}) = \sqrt{\det_{L'} E'}$ *where* $E' = \dim H'$.

Proof. Part a) follows directly from Corollary 3.8. We may choose our coordinates such that $h^i = 0$ if $1 \le i \le z$ and $h^i < 0$ if $z + 1 \le i \le z + n$. To prove b) we will show that for any $J \subset [1,\ldots,z]$ multiplication by $d\bar{z}_J$ gives an isomorphism $H_{[z+1,\ldots,z+n]}$ with $H_{J\cup[z+1,\ldots,z+n]}$. This will prove b) by Corollary 3.8 and Theorem 3.3. On the other hand the mapping is an isomorphism by the differential equation of the H_I as $h^i = 0$ if $1 \le i \le z$.

For c) we introduce a new complex structure on the real vector space $V = \mathbb{C}^g$. In the new complex structure $z_1,\ldots,z_z,\bar{z}_{z+1},\ldots,\bar{z}_{z+n},z_{z+n+1},\ldots,z_g$ are the complex coordinates. Let $\widehat{V}$ be this new complex space. Let $\widehat{H}(z,\omega)$ be the Hermitian form $\sum_{z+1\le j\le z+n} -h^j \bar{z}_j \omega_j + \sum_{j>z+n} h^j z_j \bar{\omega}_j$. Then $\widehat{H}$ and H have the same imaginary part. Now $(\alpha,\widehat{H})$ is A.-H. data for $\widehat{V}/L$ and $\widehat{H}$ is semi-positive. We claim

$$(*) \qquad\qquad H^n(X,\mathscr{L}) \cong H^0(\widehat{V}/L, \mathscr{L}(\alpha,\widehat{H})) \ .$$

If we prove this claim then part c) will follow from Theorem 2.1, its proof and Theorem 2.3 where we determined the sections of sheaves. Now for the claim let $N = [z+1,\ldots,z+n]$. Thus $H^n(X,\mathscr{L}) = H_N$ and its elements correspond to forms $\omega = f d\bar{z}_N$ in $A^0 d\bar{z}$ such that $\Delta\omega = 0$. By the usual reasoning in Kähler geometry this differential equation is equivalent to $\bar{\partial}\omega = (\bar{\partial})^*\omega = 0 \Leftrightarrow \frac{\partial f}{\partial \bar{z}_j} = 0$ if $j \notin N$ and $(\frac{\partial}{\partial \bar{z}_j})^* f = 0$ (or, rather, $-\frac{\partial}{\partial z_j} f + \pi h^j f = 0$) if $j \in N$.

The isomorphism $H^n(X,\mathscr{L}) \cong H^0(\widehat{V}/L, \mathscr{L}(\alpha,H))$ will send $\omega = f\, d\bar{z}_N$ to $g(z) = \exp(-\pi \sum_{j\in N} h^j z_j \bar{z}_j) f(z)$. The differential equations for g are $\frac{\partial g}{\partial \bar{z}_j} = 0$ if $j \notin N$ and $\frac{\partial g}{\partial z_j} = 0$ if $j \in N$; i.e., g is holomorphic on $\widehat{V}$. It only remains to check that f is in A^0 if and only if g is in $A^0(\alpha,\widehat{H})$. This is routine and we will do it one way. Assume that f is in A^0. Then

$$g(z + l) = \exp\left(-\pi \sum_{j\in N} h^j z_j \bar{z}_j - \pi\left(\sum_{j\in N} h^j(z_j\bar{l}_j + \bar{z}_j l_j + l_j\bar{l}_j)\right)\right) a_l$$

$$\times \exp(\pi H(z,l) + \frac{\pi}{2} H(l,l)) f(z)$$

$$= \alpha_l \exp\left(\pi \widehat{H}(z,l) + \frac{\pi}{2}\widehat{H}(l,l)\right) g(z) \ .$$

So g is in $A^0(\alpha,\widehat{H})$. $\qquad\qquad\qquad\qquad\qquad\qquad\qquad\qquad\square$

§3.6 Examples

Let $\mathscr{L}$ be an invertible sheaf on a complex torus $X = V/L$.

Theorem 3.10 *(Riemann-Roch). The Euler characteristic $\chi(\mathscr{L})$ of $\mathscr{L} \equiv \sum(-1)^i \dim H^i(X,\mathscr{L})$ is the intersection number $\frac{c_1(\mathscr{L})^g}{g!}$ where $g = \dim X$.*

Proof. Let $\mathscr{L} = \mathscr{L}(\alpha,H)$ be some Appel-Humbert data. Then $c_1(\mathscr{L})$ is the invariant two-form on X corresponding to the skew-symmetric form $\operatorname{Im} H = E$.

Then by linear algebra $\pm\sqrt{\det_L E}\,d\mathscr{L}_1\wedge\ldots\wedge d\mathscr{L}_{2g} = \frac{1}{g!}c_1(\mathscr{L})\wedge(g\text{ times})\wedge c_1(\mathscr{L})$. Thus H is non-degenerate if and only if $\frac{c_1(\mathscr{L})^g}{g!} = 0$.

Assume that H is degenerate we need to see that $\chi(\mathscr{L}) = 0$. This case follows from Theorem 3.9 a) and b). Because $\sum(-1)^i H^i(K^0(\mathscr{L}), \mathcal{O}_{K^0(\mathscr{L})}) = 0$ as the cohomology of the structure sheaf is an exterior algebra. If H is non-degenerate then $z = 0$ and $H^i(X, \mathscr{L})$ is non-zero when $i = n$ and its dimension is $\pm\sqrt{\det_L E}$. Thus we need only check that $\pm = (-1)^n$. This is a question in linear algebra which we don't do. (Hint: if H' is a positive definite form on V when $(\mathrm{Im}\, H')^g$ is positive). $\qquad\square$

A frequently used special case is

Corollary 3.11. *If D is an ample divisor on an abelian variety X then*

   a) $\dim\Gamma(X, \mathcal{O}_X(D)) =$ the intersection number $\frac{1}{g!}D^g$ and

   b) $H^i(X, \mathcal{O}_X(D)) = 0$ if $i > 0$.

Another special case of Theorem 3.9 is

Corollary 3.12. *If $\mathscr{L}$ is in $\mathrm{Pic}^0(X)$ but $\mathscr{L} \neq \mathcal{O}_X$, then $H^i(X, \mathscr{L}) = 0$ for all i.*

Let $\mathscr{P}$ be the Poincaré sheaf on $X \times X\widehat{\ }$ where $X\widehat{\ }$ is dual complex torus.

Corollary 3.13. *The only non-zero cohomology group of $\mathscr{P}$ is $H^g(X \times X\widehat{\ }, \mathscr{P})$ which has dimension one.*

Proof. We left this as an exercise (Hint: prove that $\chi(\mathscr{P}) = \pm 1$ and that the tangent space of $X(\dim g)$ is an isotropic subspace of $H(\mathscr{P})$). $\qquad\square$

Next we want to make some applications to families of cohomology groups. Let $f : X \to S$ be a smooth proper morphism of connected analytic spaces. Assume that the fibers $X_s = f^{-1}(s)$ are abelian varieties. Let $\mathscr{L}$ be an invertible sheaf of X.

Theorem 3.14. *If for all s in S the sheaf $\mathscr{L}|_{X_s}$ is ample on X_s, then*

   a) $f_*\mathscr{L}$ is a locally free sheaf on S and for each s, $(f_*\mathscr{L})_s \overset{\sim}{\underset{\approx}{\to}} \Gamma(X_s, \mathscr{L}|X_s)$ is an isomorphism.

   b) $R^i f_*\mathscr{L} = 0$ if $i > 0$.

Proof. Follows from Corollary 3.11 by proper flat base extension. (Actually a) can be proven directly for using the Fourier expansions of sections of $\mathscr{L}$ along the fibers.) $\qquad\square$

We will need to know one calculation of higher direct images for the Poincaré sheaf $\mathscr{P}$.

Theorem 3.15. $R^g \pi_{X^\wedge *}(\mathscr{P})$ *is the one-dimensional sky-scraper sheaf situated at the zero point 0 of* $X^\wedge$ *and the other higher direct images are zero.*

Proof. By proper flat base extension for $\pi_{X^\wedge} : X \times X^\wedge \to X^\wedge$ the Corollary 3.12 implies that support $(R^i \pi_{X^\wedge *}(\mathscr{P})) \subset \{0\}$ for all i. Hence the Leray spectral sequence gives an isomorphism

$$R^i \pi_{X^\wedge *}(\mathscr{P})_{(0)} = \Gamma(X^\wedge, R^i \pi_{X^\wedge *}(\mathscr{P})) \cong H^i(X \times X^\wedge, \mathscr{P}) \ .$$

Thus this result follows from 3.13. $\qquad\square$

Chapter 4. Groups Acting
on Complete Linear Systems

§ 4.1 Geometric Background

Let $\mathscr{L}$ be a very ample invertible sheaf on an abelian variety X. Then we have a projective embedding of X in $\mathbb{P}^n$ by the complete linear system $|D|$ of sections of $\mathscr{L}$.

Lemma 4.1. *Translation $T_x : X \to X$ by a point x of X extends to a projective transformation of $\mathbb{P}^n$ if and only if x is contained in the finite group $K(\mathscr{L}) = \{x \in X | T_x^* \mathscr{L} \approx \mathscr{L}\}$.*

Proof. We have already seen in Lemma 1.10 that $K(\mathscr{L})$ is finite. Assume that T_x extends to a projective transformation. Then for any E in $|D|$ the divisor $E - x$ is linearly equivalent to E. As $\mathscr{L} \approx \mathscr{O}_X(E)$ and $T_x^* \mathscr{L} \approx \mathscr{O}_X(E - x)$ we must have x in $K(\mathscr{L})$. Conversely if $\alpha : T_x^* \mathscr{L} \xrightarrow{\approx} \mathscr{L}$ is an isomorphism, then the global effect of α defines an isomorphism $\mathbb{P}^n \to \mathbb{P}^n$ which extends the action of T_x on X. $\qquad\square$

As the above extension is clearly unique because X spans $\mathbb{P}^n$, we have a projective representation of the finite abelian group $K(\mathscr{L})$ on $\mathbb{P}^n$. It will shortly turn out that $\mathbb{P}^n$ is a very simple irreducible representation of $K(\mathscr{L})$ but the purpose of this section is to lay the groundwork for this presentation.

Let $\rho : G \to \mathrm{PGL}(n)$ be a projective representation of a group G. Then we have an exact commutative diagram

$$
\begin{array}{ccccccccc}
1 & \longrightarrow & \mathbb{C}^* & \longrightarrow & H & \longrightarrow & G & \longrightarrow & 0 \\
 & & \| & & \downarrow{\rho'} & & \downarrow{\rho} & & \\
1 & \longrightarrow & \mathbb{C}^* & \longrightarrow & \mathrm{GL}(n+1) & \xrightarrow{\sigma} & \mathrm{PGL}(n) & \longrightarrow & 1
\end{array}
$$

where H is the fiber product of σ and ρ. Thus our projective representation ρ is determined by the ordinary representation ρ of the central extension H of G by $\mathbb{C}^*$.

Now let $\mathscr{L}$ be an invertible sheaf on X, we want to construct the natural central extension $H(\mathscr{L})$ of $K(\mathscr{L})$ by $\mathbb{C}^*$ together with a representation of $H(\mathscr{L})$ on $\Gamma(X, \mathscr{L})$ which will give above projective representation of $K(\mathscr{L})$ on $\mathbb{P}^n$ when $\mathscr{L}$ is very ample. This group $H(\mathscr{L})$ is called the theta (or Heisenberg) group of $\mathscr{L}$.

By definition an element of $H(\mathscr{L})$ is a pair (x, α) where x is a point of $K(\mathscr{L})$ and α is an isomorphism $\alpha : T_x^* \mathscr{L} \xrightarrow{\approx} \mathscr{L}$. The product $(y, \beta) * (x, \alpha) = (x + y, \beta \circ T_y^* \alpha)$ where $\beta \circ T_y^*(\alpha)$ is the composition $T_{x+y}^* \mathscr{L} = T_y^*(T_x^* \mathscr{L}) \xrightarrow{T_y^*(\alpha)} T_y^* \mathscr{L} \xrightarrow{\beta} \mathscr{L}$. One checks that $H(\mathscr{L})$ is a group with identiy $(0, 1)$. Furthermore $\mathbb{C}^*$ is a subgroup of $H(\mathscr{L})$ if we identify a complex number k with $(0, k)$ and we have a central extension $1 \to \mathbb{C}^* \to H(\mathscr{L}) \to K(\mathscr{L}) \to 0$. To get the corresponding representation on $\Gamma(X, \mathscr{L})$ for (x, α) in $H(\mathscr{L})$ and a section σ of $\mathscr{L}$, define $(x, \alpha) \cdot \sigma$ to be $\alpha(T_x^*(\sigma))$. From the definition this gives a representation of $H(\mathscr{L})$ in which $\mathbb{C}^*$ acts by multiplication.

To study the non-commutativity of $H(\mathscr{L})$, we introduce the form $e_{\mathscr{L}} : K(\mathscr{L}) \times K(\mathscr{L}) \to \mathbb{C}^*$ which is defined by the commutators in $H(\mathscr{L})$. Explicitly $e_{\mathscr{L}}(x, y) = (x, \alpha)(y, \beta)(x, \alpha)^{-1}(y, \beta)^{-1}$ where (x, α) and (y, β) are any elements of $H(\mathscr{L})$ lying over x and y in $K(\mathscr{L})$. Next we compute the form $e_{\mathscr{L}}$.

Let (α, H) be A.-H. data for $\mathscr{L}$ and $E = \operatorname{Im} H$ where $X = V/L$ as usual. Recall that $K(\mathscr{L}) = L^{\perp}/L$ where $L^{\perp} = \{v \in V | E(v, l) \in \mathbb{Z} \text{ for all } l \text{ in } L\}$. Thus if m and n are elements of $L^{\perp}$, we want

Lemma 4.2. $e_{\mathscr{L}}(\bar{m}, \bar{n}) = e^{-2\pi i E(m, n)}$.

Proof. We will first describe the transformation $(\bar{m}, A)$ in $H(\mathscr{L})$ lifting the element $\bar{m}$ in $K(\mathscr{L})$. It is convenient to think of the $T_{\bar{m}}^*$: homomorphism $A' : \mathscr{L} \to \mathscr{L}$ associated to A, which is gotten by $A \circ T_{\bar{m}}^*$. Recall that $\mathscr{L}$ is the sheaf which assigns to each L invariant open subset U of V, the holomorphic funtions f on U such that $f(u + l) = \alpha_l e^{\pi H(u, l) + \frac{\pi}{2} H(l, l)} f(u)$. We claim that $A'(f)(u) = e^{-\pi H(u, m) - \frac{\pi}{2} H(m, m)} f(u + m)$ is such an operation. The point is that A' takes $\mathscr{L}$ into $\mathscr{L}$ and is a T_m^*-homomorphism almost by definition. To show that A' takes $\mathscr{L}$ into $\mathscr{L}$ note that if f is a section of $\mathscr{L}$ over u, then (after some calculations) $A'(f)(u + l) = e^{2\pi i E(m, l)} \alpha_l e^{\pi H(u, l) + \frac{\pi}{2} H(l, l)} A'(f)(u)$. Hence as $E(m, l)$ is always an integer $A'(f)$ is a section of $\mathscr{L}$ over $U - m$.

Next let B' be an operator with n instead of m. We need to find the constant $\lambda = e_{\mathscr{L}}(m, n)$ such that $A' \cdot B' = \lambda B' \cdot A'$. Now $A' \cdot B'(f)(u) = e^{-\pi H(m, n)}$ (terms symmetric in m and n). Therefore $\lambda = e^{-2\pi i E(m, n)}$. $\qquad \square$

Corollary 4.3. *If $\mathscr{L}$ is non-degenerate then the center of $H(\mathscr{L})$ is $\mathbb{C}^*$.*

Proof. Given a non-zero element $\bar{m}$ of $K(\mathscr{L})$ we need to find another element $\bar{n}$ of $K(\mathscr{L})$ such that $e_{\mathscr{L}}(\bar{m}, \bar{n}) \neq 1$. In other words such that $E(m, n)$ is not an integer. By assumption we know that the form E is non-degenerate. We need to see that if $E(m, n)$ is an integer for all n in M then M is in L. This is easy. Let $l_1, \ldots, l_{2g}$ be a basis for L. Then the dual basis is a basis for M. Hence M is contained in the span of the original basis (double dual basis) which is L. $\square$

Exercise 1. Define a natural structure of complex analytic group on $H(\mathscr{L})$ such that the sequence

$$1 \longrightarrow \mathbb{C}^* \longrightarrow H(\mathscr{L}) \longrightarrow K(\mathscr{L}) \longrightarrow 0$$

is an exact sequence of complex analytic group.

§ 4.2 Representations of the Theta Group

The theory of projective representation of finite abelian groups is an exercise for a college student. We will present a special case where we can be very explicit.

Let $H(\mathscr{L})$ be the theta group for a nondegenerate sheaf $\mathscr{L}$ on an abelian variety. Let $\pi : H(\mathscr{L}) \to K(\mathscr{L})$ be the projection. Take a subgroup $A(\mathscr{L})$ of $K(\mathscr{L})$ which is maximal with respect to the requirement that $\pi^{-1}A(\mathscr{L})$ is abelian. Then the form $e_{\mathscr{L}}$ defines a homomorphism

$$\ell : K(\mathscr{L}) \longrightarrow \mathrm{Hom}_{\mathbf{Z}}\left(A(\mathscr{L}), \mathbb{C}^*\right)$$

by $\ell(k)(a) = e_{\mathscr{L}}(a, k)$. The kernel of ℓ is exactly $A(\mathscr{L})$ by maximality of $A(\mathscr{L})$. By Corollary 4.3 the similar mapping $\ell' : K(\mathscr{L}) \to \mathrm{Hom}_{\mathbf{Z}}(K(\mathscr{L}), \mathbb{C}^*)$ is injective. Hence it is an isomorphism because the dual group has the same order. Therefore ℓ is a surjection because $\mathrm{Hom}_{\mathbf{Z}}(K(\mathscr{L}), \mathbb{C}^*) \to \mathrm{Hom}_{\mathbf{Z}}(A(\mathscr{L}), \mathbb{C}^*)$ is surjective. In summary we have an exact sequence

$$0 \longrightarrow A(\mathscr{L}) \longrightarrow K(\mathscr{L}) \longrightarrow \mathrm{Hom}_{\mathbf{Z}}\left(A(\mathscr{L}), \mathbb{C}^*\right) \longrightarrow 0$$

and hence $\#A(\mathscr{L})^2 = \#K(\mathscr{L})$.

We want to consider an irreducible representation V of $H(\mathscr{L})$ on which $\mathbb{C}^*$ acts by multiplication. As $\pi^{-1}A(\mathscr{L})$ abelian we may find a non-zero eigenvector v_χ with eigenvalue a character χ of $\pi^{-1}A(\mathscr{L})$ which is good in sense that it is one on $\mathbb{C}$. Two such good characters defer by multiplication π^* of a character of $A(\mathscr{L})$. Now let h be an element of $H(\mathscr{L})$. So $h \cdot v_\chi$ is also an eigenvector with eigenvalue $\chi \pi^* \ell(\pi h)$ because $ahv_\chi = e_{\mathscr{L}}(a, h)h \cdot av_\chi = \chi(a)e_{\mathscr{L}}(\pi a, \pi h)h \cdot v_\chi$. Therefore χ could be an arbitrary good character of $\pi^{-1}A(\mathscr{L})$. As the line spanned by hv_χ depends only on the coset of k in $H(\mathscr{L})/A(\mathscr{L})$, we have $\bigoplus_{k \in H(\mathscr{L})/A(\mathscr{L})} \mathbb{C}k \cdot v_\chi$ is a $H(\mathscr{L})$-invariant subspace of V. By reducibility we have $V = \bigoplus_{k \in H(\mathscr{L})/\pi^{-1}A(\mathscr{L})} \mathbb{C}k \cdot v_\chi$. Clearly the action V is simply determined by our implicit choice of coset representatives but nothing else because if $l \cdot k = k'a$ where k and k' are coset representatives and a is in $\pi^{-1}A(\mathscr{L})$, then $l(k \cdot v_\chi) = \chi(a) \cdot k' \cdot v_\chi$.

Theorem 4.4. *The theta group $H(\mathscr{L})$ has a unique upto isomorphism representation Γ on which $\mathbb{C}^*$ acts by multiplication. Furthermore* $\dim \Gamma = \sqrt{\#K(\mathscr{L})}$.

Proof. We have seen the uniqueness. Conversely if one defines V as above, one need only know that there is a good character χ of $\pi^{-1}A$: i.e. a splitting

$$1 \longrightarrow \mathbb{C}^* \underset{}{\overset{\chi}{\rightleftarrows}} \pi^{-1}A(\mathscr{L}) \longrightarrow A(\mathscr{L})$$

of this sequence of abelian groups (as an extension of a finite group by a divisible group splits!). For the dimension count note that $\dim V = \#H(\mathscr{L})/\pi^{-1}A(\mathscr{L}) = \#K(\mathscr{L})/A(\mathscr{L}) = \sqrt{\#K(\mathscr{L})}$. $\qquad\qquad\square$

Theorem 4.5. *Let i be the index of $\mathscr{L}$. Then the natural representation of $H(\mathscr{L})$ on $H^i(X, \mathscr{L})$ is isomorphic to Γ.*

Proof. By Theorem 3.9 $\dim H^i(X, \mathscr{L}) = \det_L E = \sqrt{K(\mathscr{L})}$ by Lemma 1.10. The natural action of (x, α) in $H(\mathscr{L})$ on $H^i(X, \mathscr{L})$ is $(x, \alpha) \cdot \sigma = \alpha \circ T_x^*(\sigma)$ and so $\mathbb{C}^*$ acts by multiplication. Therefore by Theorem 4.4 this theorem is true. $\qquad\qquad\square$

In the rest of this section we will describe the standard form of the representation Γ. First of all having a good character χ of $\pi^{-1}A(\mathscr{L})$ is the same as having a subgroup $A'(\mathscr{L})$ of $\pi^{-1}A(\mathscr{L})$ $(= \operatorname{Ker}\chi)$ such that π induces an isomorphism $A'(\mathscr{L}) \xrightarrow{\approx} A(\mathscr{L})$. Such a subgroup is called a level subgroup. Now assume that we have another subgroup $B(\mathscr{L})$ with the same properties as $A(\mathscr{L})$ and lifting $A'(\mathscr{L})$ and $B'(\mathscr{L})$ of $A(\mathscr{L})$ and $B(\mathscr{L})$ and $K(\mathscr{L}) = A(\mathscr{L}) \oplus B(\mathscr{L})$. The choice of such subgroups $A'(\mathscr{L})$ and $B'(\mathscr{L})$ is called a decomposition of $H(\mathscr{L})$.

We can be very explicit about the theta group $H(\mathscr{L})$. Any of its elements can be written uniquely as $\lambda \cdot a \cdot b$ where $\lambda \in \mathbb{C}$, $a \in A'(\mathscr{L})$ and $b \in B'(\mathscr{L})$. Thus $H(\mathscr{L}) \approx \mathbb{C} \times A'(\mathscr{L}) \times B'(\mathscr{L}) \approx \mathbb{C} \times A(\mathscr{L}) \times B(\mathscr{L})$ as set. The product in $H(\mathscr{L})$ is given by

$$(\lambda, a, b) \cdot (\lambda', a', b') = (\lambda\lambda' e_{\mathscr{L}}(b, a'), a + a', b + b') .$$

We can represent Γ as functions on $B(\mathscr{L})$. The group action is given by

$$((\lambda, a, b) \cdot f)(c) = \lambda e_{\mathscr{L}}(a, c)^{-1} f(c + b) \ .$$

It is trivial to check that this is such an irreducible representation. Call this the standard method for Γ. Lastly we need in this situation $B(\mathscr{L})$ is identified with the character group of $A(\mathscr{L})$ by the form $e_{\mathscr{L}}$.

The analytic theory of theta functions involve the construction of an explicit isomorphism of $H(\mathscr{L})$-modules

$$\rho_{\mathscr{L}} : \Gamma \xrightarrow{\ \approx\ } \Gamma(X, \mathscr{L})$$

where $\mathscr{L}$ is an ample sheaf on an abelian variety. The algebraic theory says that $\rho_{\mathscr{L}}$ is determined upto constant multiple (apply Schur's lemma to these irreducible representation). The analytic theory allows a determination of this multiple.

Next we will note a lemma of Mumford which uses the irreducibility.

Lemma 4.6. *Let* $\mathscr{L} \approx \mathscr{N} \otimes \mathscr{M}$ *be an ample sheaf on an abelian variety such that* $\mathscr{N}$ *and* $\mathscr{M}$ *have non-zero sections. Then*

$$\Gamma(X, \mathscr{L}) = \sum \Gamma(X, \mathscr{N}') \cdot \Gamma(X, \mathscr{M}')$$

where $\mathscr{L} = \mathscr{N}' \otimes \mathscr{M}'$ *and* $\mathscr{N}'$ ($\mathscr{M}'$) *is topologically equivalent to* $\mathscr{N}$ ($\mathscr{M}$).

Proof. The left side of the inequality is a non-zero subspace. By Theorem 4.5 it will suffice to show that it is invariant under the action of $H(\mathscr{L})$ because $\Gamma(X, \mathscr{L})$ is irreducible. Let $\alpha : T_x^* \mathscr{L} \xrightarrow{\ \approx\ } \mathscr{L}$ be an isomorphism in $H(\mathscr{L})$. By definition $\alpha(\Gamma(X, \mathscr{N}')\Gamma(X, \mathscr{M}')) = \alpha(\Gamma(X, T_x^* \mathscr{N}')) \cdot \Gamma(X, T_* \mathscr{M}') = \Gamma(X, \mathscr{N}'') \cdot \Gamma(X, \mathscr{M}'')$ where $\mathscr{N}'' = T_x^* \mathscr{N}'$ and $\mathscr{M}'' = T_x^* \mathscr{M}'$. As $\mathscr{N}'' \otimes \mathscr{M}'' \approx \mathscr{L}$ and $\mathscr{N}''$ ($\mathscr{M}''$) is equivalent to $\mathscr{N}$ ($\mathscr{M}$), we are done. $\square$

§ 4.3 The Hermitian Structure on $\Gamma(X, \mathscr{L})$

Let $\mathscr{L}$ be an invertible sheaf on an abelian variety X. Let $X = V/L$ and $\mathscr{L} = \mathscr{L}(\alpha, H)$ for some A.-H. data (α, H). In the proof of Theorem 2.1 we have introduced a canonical Hermitian inner product in $\mathscr{L}(\alpha, H)$ by defining

$$(\sigma, \tau)_x = e^{-\pi H(\tilde{x}, \tilde{x})} \sigma(\tilde{x}) \overline{\tau(\tilde{x})}$$

where $\tilde{x}$ is a point of V over a point x in X and σ and τ are sections of $\mathscr{L}(\alpha, H)$ defined around x. Thus $\mathscr{L}(\alpha, H)$ has a natural Hermitian invertible sheaf structure.

Such sheaves have a remarkable property.

Lemma 4.7. *Let α and β be two Hermitian isomorphisms $\mathscr{L}_1 \rightrightarrows \mathscr{L}_2$ between two invertible sheaves with Hermitian structure over a connected complex manifold Y. Then α and β differ by multiplication by a unitary complex number.*

Proof. Consider $\gamma = \beta^{-1}\alpha$. It is a Hermitian endomorphism of $\mathscr{L}_1$ which is nowhere zero. Let σ be a nowhere vanishing local section of $\mathscr{L}_1$. Then $\gamma(\sigma)$ has the same property and the same length as σ. Therefore the ratio $\gamma(\sigma)/\sigma$ is a holomorphic function which always has absolute value 1. Hence it is a constant u (locally) where $|u| = 1$. Thus $u = \beta^{-1}\alpha$ or rather $u\beta = \alpha$ as Y is connected. $\qquad\Box$

This gives us a compact form $H_c(\mathscr{L})$ of the theta group $H(\mathscr{L})$. The element (x, α) of $H(\mathscr{L})$ are in $H_c(\mathscr{L})$ if the isomorphism $\alpha : T_x^*\mathscr{L} \xrightarrow{\approx} \mathscr{L}$ is Hermitian when we get $T_x^*\mathscr{L}$ the Hermitian structure gotten by pulling back the natural one on $\mathscr{L}$. One can easily check that $H_c(\mathscr{L})$ is a subgroup of $H(\mathscr{L})$. Clearly we have a complex

$$(*) \qquad 1 \longrightarrow U(1) \longrightarrow H_c(\mathscr{L}) \longrightarrow K(\mathscr{L}) \longrightarrow 0 \,.$$

This is exact on the left by Lemma 4.6. To check exactness on the right we need to know that for any x in $K(\mathscr{L})$, we can find (x, α) with α Hermitian. We will do this explicitly by noting the transformation A' of Lemma 4.2 as Hermitian by verifying of the definitions.

If $\mathscr{L}$ is ample then we have a Hermitian inner product of $\Gamma(X, \mathscr{L})$ given by

$$(\sigma, \tau) = \int_{V/L} \sigma(v)\bar{\tau}(v)f(v)dv$$

where the invariant measure dv on V/L is normalized such that $\int_{V/L} dv = 1$ where $f(v) = e^{-\pi H(v,v)}$ is the metric in $\mathscr{L}(\alpha, H)$ introduced in the proof of Theorem 2.1. By definition the action of the compact theta group $H_c(\mathscr{L})$ preserves these inner products on $\Gamma(X, \mathscr{L})$.

Let $A'(\mathscr{L})$ and $B'(\mathscr{L})$ be a decomposition of $H(\mathscr{L})$. The decomposition is unitary if $A'(\mathscr{L})$ and $B'(\mathscr{L})$ are contained in $H_c(\mathscr{L})$. In the last section we have introduced an identification of the representation Γ with the functions on $B'(\mathscr{L}) \approx B(\mathscr{L})$ in $K(\mathscr{L})$. We also may introduce a standard Hermitian structure on Γ such that the action of $H_c(\mathscr{L})$ is unitary. The inner product (f, g) of two functions f and g on $B(\mathscr{L})$ is $\sum_{b \in B(\mathscr{L})} f(b)\bar{g}(a)$. Thus the basis of delta functions $\{\delta_b\}$ is a unitary basis of Γ. It is automatic from the definition that this inner product is $H_c(\mathscr{L}) = U(1) \times A'(\mathscr{L}) \times B'(\mathscr{L})$ invariant.

In the above situation we may measure the length of an analytic theory $\rho_{\mathscr{L}} : \mathbb{C}(B(\mathscr{L})) \xrightarrow{\approx} \Gamma(X, \mathscr{L})$ of the theta functions. As ρ_α is $H(\mathscr{L})$-invariant.

as we are dealing with irreducible representation $\rho_{\mathscr{L}}$ multiplies the length of vectors by a constant. Thus if we define $\|\rho_{\mathscr{L}}\| = \sqrt{(\rho_{\mathscr{L}}\delta_0, \rho_{\mathscr{L}}\delta_0)}$. Hence for any functions f and g on $B(\mathscr{L})$ we have $(\rho_{\mathscr{L}}f, \rho_{\mathscr{L}}g) = \|\rho_{\mathscr{L}}\|^2(f, g)$. .

§ 4.4 The Isogeny Theorem up to a Constant

Let $f : X \to Y$ be an isogeny of abelian varieties. Let M be an ample invertible sheaf on Y. Then $\mathscr{L} = f^*\mathscr{M}$ is ample on X. We intend to describe the pull-back mapping $f^* : \Gamma(Y, \mathscr{M}) \to \Gamma(X, \mathscr{L})$ using a theory of theta functions of $\mathscr{M}$ and $\mathscr{L}$. If we have a compatible (to be defined) decompositions $(A'(\mathscr{L}), B'(\mathscr{L}))$ of $H(\mathscr{L})$ and $A'(\mathscr{M}), B'(\mathscr{M}))$ of $H(\mathscr{M})$.

We will state the objective first. Assume that we have compatibility and isomorphisms $\rho_{\mathscr{L}} : \mathbb{C}[B(\mathscr{L})] \xrightarrow{\approx} \Gamma(X, \mathscr{L})$ and $\rho_{\mathscr{M}} : \mathbb{C}[B(\mathscr{M})] \xrightarrow{\approx} \Gamma(Y, \mathscr{M})$ which are $H(\mathscr{L})$ and $H(\mathscr{M})$-module homomorphisms.

Theorem 4.7. $\rho_{\mathscr{L}}^{-1} f^* \rho_{\mathscr{M}}(\delta_b) = $ constant $(\sum_{\substack{b' \in B(\mathscr{L}) \\ f(b')=b}} \delta_{b'})$ *for all b in* $B(\mathscr{M})$.

This remarkably simple formula is the reason that the theory of isogenies is simple. Basically it says that the algebra of f^* is determined upto constant by the geometry of f. The analytic theory of the next chapter will determine the constant in the formula.

To understand the meaning of compatibility we need to understand the relationship between the two theta groups $H(\mathscr{L})$ and $H(\mathscr{M})$.

We may assume that $X = V/L$ and $Y = V/M$ where the lattice M contains L and $f(v + L) = v + M$. Also assume that $\mathscr{M}$ is given by A.-H. data (α, H). Then $\mathscr{L}$ is given by $(\alpha|_L, H)$. We have inclusions $L^\perp \supset M^\perp \supset M \supset L$. Hence we have a diagram

$$K(\mathscr{L}) = L^\perp/L \supset M^\perp/L$$
$$\downarrow f$$
$$M^\perp/M = K(\mathscr{M})$$

where $M^\perp/L = \{x \in L^\perp/L \,|\, e^{2\pi i E(x,m)} = 1 \text{ for all } m \text{ in } M/L = \operatorname{Ker} f\}$.

Now we have a natural homomorphism $\operatorname{Ker}(f) \hookrightarrow H(\mathscr{L})$ which sends a coset m of M into the transformation represented by

$$f(v) \longrightarrow \alpha(m)^{-1} e^{-\pi H(v,m) \frac{\pi}{2} H(m,m)} f(v + m) \ .$$

Trivially this is a homomorphism and $\Gamma(Y, \mathscr{M})$ is the subspace of $\operatorname{Ker}(f)$-invariants in $\Gamma(X, \mathscr{L})$. By Lemma 4.2 the centralizer N of $\operatorname{Ker}(f)$ is the inverse image of $M^\perp/L$ under $H(\mathscr{L}) \to K(\mathscr{L})$. There is a natural surjective homomorphism $\beta : N \to H(\mathscr{M})$ with kernel $= \operatorname{Ker}(f)$ given by sending a transformation into the same formula by thinking of it operating on different sheaves $\mathscr{L}(\alpha|_L, H)$ and $\mathscr{L}(\alpha, H)$.

Compatibility means that

a) $\beta(A'(\mathscr{L}) \cap N) = A'(\mathscr{M})$,

b) $\beta(B'(\mathscr{L}) \cap N) = B'(\mathscr{M})$, and

c) $\mathrm{Ker}(f) = (\mathrm{Ker}(f) \cap A'(\mathscr{L})) \oplus (\mathrm{Ker}(f) \cap B'(\mathscr{L}))$.

Now we are ready for the

Proof of Theorem 4.7. By Theorem 4.5, $\Gamma(Y, \mathscr{M})$ is an irreducible representation of $H(\mathscr{M})$. Now $f^* : \Gamma(Y, \mathscr{M}) \to \Gamma(X, \mathscr{L})$ is β-linear (i.e. $f^*(\beta(x) \cdot \sigma) = x \cdot f^*(\sigma)$ for x in N). As β is surjective it follows that f^* is uniquely determined upto to constant multiple as an β-linear injection. To prove this theorem just note that $\eta_{\mathscr{L}}^{-1} f^* \eta_{\mathscr{M}}$ is also a β-linear injection by an exercise using the actions on $\mathbb{C}[B(\mathscr{L})]$ and $\mathbb{C}[B(\mathscr{M})]$ and the compatibility conditions. $\qquad\square$

Exercise 1. Finish the proof of the theorem with complete details.

Chapter 5. Theta Functions

§ 5.1 Canonical Decompositions and Bases

Let L be a free abelian group of finite rank with a non-degenerate integral-valued skew-symmetric form $E : L \times L \to \mathbb{Z}$. In this section we will develop the structure of the symplectic lattice L.

An isotropic subgroup is a subgroup A of L such that $E(a_1, a_2) = 0$ for all a_1 and a_2 in A.

Lemma 5.1. *There exists two isotropic subgroups A and B of L such that $A \oplus B = L$.*

Proof. If $L \neq 0$ then there are elements a and b of L such that $E(a, b) \equiv m$ is positive and minimal. Thus the subgroup of values $E(a, l)$ (or $E(l, b)$) is generated by m. For any l in L we may write $l = l' + (E(a, l)/m)b - (E(b, l)/m)a$ where $E(l, a) = 0$. If $(a, b)^\perp$ denote the subspace of all such l, we have $\mathbb{Z}a \oplus \mathbb{Z}b \oplus (a, b)^\perp$. Here $(a, b)^\perp$ is non-degenerate for E and we may repeat the construction if necessary. Therefore we have a basis $a_1, b_1, a_2, b_2, \ldots, a_g, b_g$ of L such that all these elements are perpendicular except for all a_i and b_i. For this lemma we may take $A = \oplus \mathbb{Z}a_i$ and $B = \oplus \mathbb{Z}b_i$. $\qquad\square$

So the rank L is even. We can even find a "canonical" basis $a_1, b_1, a_2, b_2, \ldots, a_g, b_g$ such that they are all perpendicular except a_i and b_i and $E(a_1, b_1)$ $|E(a_2, b_2)| \ldots |E(a_g, b_g)$ as positive integers.

Lemma 5.2. *Given a decomposition $A \oplus B = L$ as in Lemma 1 we may find bases $a_1, \ldots, a_g$ of A and $b_1, \ldots, b_g$ of B which form a canonical basis.*

Proof. The form E gives an homomorphism $\varphi : B \to \mathrm{Hom}_{\mathbb{Z}}(A, \mathbb{Z})$ which is an isomorphism upto torsion. By the fundamental theorem of abelian groups we may find a basis $b_1, \ldots, b_g$ of B and $\widehat{a_1}, \ldots, \widehat{a_g}$ of $\mathrm{Hom}_{\mathbb{Z}}(A, \mathbb{Z})$ such that the matrix giving φ is diagonal and positive with successively division of the coefficients. Now let $a_1, \ldots, a_g$ be the dual basis of A. Then we have solved the problem. $\square$

Exercise 1. Find a canonical basis of $\mathbb{Z}^4$ where E is given by the matrix

$$\begin{bmatrix} 0 & 2 & -3 & 0 \\ -2 & 0 & -5 & -4 \\ 3 & 5 & 0 & -2 \\ 0 & 4 & 2 & 0 \end{bmatrix} .$$

Exercise 2. Using the existence of a canonical basis show that the proof of Lemma 5.1 actually produces one.

Exercise 3. Show that the integers $\{E(a_i, b_i)\}$ depend only on L and E. (Hint: What are the elementary divisors of matrix representing E?).

§ 5.2 The Theta Function

Let X be an abelian variety V/L with polarization H. Let $A \oplus B$ be a decomposition of the lattice L with respect to the skew-symmetric form $E = \mathrm{Im}\, H$. By the proof of Lemma 1.6 we have a unique multiplier α on L such that $\alpha|A = \alpha|B = 1$ such that (α, H) is A.-H. data. The sheaf $\mathcal{L} = \mathcal{L}(\alpha, H)$ is called excellent with respect to the decomposition $A \oplus B = L$, or just excellent if there is no confusion. The whole theory of theta function revolves around the special properties of excellent sheaves.

Consider $L^{\hat{}} = \{v \in V | E(v, l) \in \mathbb{Z} \text{ for all } l \text{ in } L\}$. We have $L^{\hat{}} = A^{\hat{}} \oplus B^{\hat{}}$ where $A = L^{\hat{}} \cap A \otimes \mathbb{R}$ and $B = L^{\hat{}} \cap B \otimes \mathbb{R}$. Thus $K(\mathcal{L}) = L^{\hat{}}/L$ is the direct sum $A^{\hat{}}/A \oplus B^{\hat{}}/B$. Let $A(\mathcal{L}) = A^{\hat{}}/A$ and $B(\mathcal{L}) = B^{\hat{}}/B$. One may check that $A(\mathcal{L})$ and $B(\mathcal{L})$ are maximal isotropic subgroups of $K(\mathcal{L})$.

Lemma 5.3. *There is a canonical decomposition $A'(\mathcal{L})$ and $B'(\mathcal{L})$ of $H(\mathcal{L})$ which refines $A(\mathcal{L})$ and $B(\mathcal{L})$. Furthermore $A'(\mathcal{L})$ and $B'(\mathcal{L})$ lie in $H_c(\mathcal{L})$.*

Proof. We need to define section σ_A and σ_B of $H_c(\mathcal{L}) \to K(\mathcal{L})$ over $A(\mathcal{L})$ and $B(\mathcal{L})$. If a is in $A(\mathcal{L})$ then let $\sigma_A(a)$ be given by the transformation $f(v) \to \exp(-\pi H(v, \tilde{a})\frac{\pi}{2}H(\tilde{a}, \tilde{a}))f(v + \tilde{a})$ where $\tilde{a}$ in V lies over a. We have already noted that $\sigma_A(a) \in H_c(\mathcal{L})$. One must check that σ_A is a homomorphism. This follows from a routine calculation as E is zero on $A^{\hat{}}$. Furthermore it does not depend on the choice of $\tilde{a}$. Then section σ_B is defined in the same way. $\square$

The central result is

Lemma 5.4. *There is a canonical $H(\mathscr{L})$-module isomorphism*

$$\rho_{\mathscr{L}} : \mathbb{C}(B(\mathscr{L})) \longrightarrow \Gamma(X, \mathscr{L}) \ .$$

Proof. $\rho_{\mathscr{L}}(\delta_0)$ must be a $A'(\mathscr{L})$-invariant section of $\mathscr{L}$. Then $\rho_{\mathscr{L}}(\delta_b) = \sigma_B(b)(\rho_{\mathscr{L}}(\delta_0))$. Thus it will suffice to determine $\rho_{\mathscr{L}}(\delta_0)$ explicitly. We return to the proof of Theorem 2.13. Set $U = A$, $U' = B$ and $\lambda = 0$. Let $\rho_{\mathscr{L}}(\delta_0) = e^{\pi/2B(v,v)}(\sum_{b\in B} e^{-\pi i b\widehat{\ }(b)+2\pi i b\widehat{\ }(v)})$. By the theorem this defines a non-zero section of $\mathscr{L}$. We must check that it is $A'(\mathscr{L})$-invariant. In other words that it is a section of the excellent sheaf on $V/A\widehat{\ } \oplus B$ with polarization H, but this follows from the theorem in that case. $\square$

The expression $\sum_{b\in B} e^{-\pi i b\widehat{\ }(b)+2\pi i b\widehat{\ }(v)}$ is called the theta function. It is useful to have an explicit expression for $\rho_{\mathscr{L}}(\delta_b)$. Note that the constant term of this Fourier series is 1.

Corollary 5.5. *For c in $B(\mathscr{L})$ represented by an element $\tilde{c}$ of $B\widehat{\ }$ we have*

$$\rho_{\mathscr{L}}(\delta_c) = e^{\pi/2S(v,v)}\left(\sum_{b\in B} e^{-\pi i(b-\tilde{c})\widehat{\ }(b-\tilde{c})+2\pi i(b+\tilde{c})\widehat{\ }(v)}\right) \ .$$

Proof.

$$\rho_{\mathscr{L}}(\delta_c)(v) = \rho_{\mathscr{L}}((0,0,-c)^*\delta_0) = \sigma_B(-c) \cdot \rho_{\mathscr{L}}(\delta_0)(v)$$

$$= e^{-\pi H(v,-\tilde{c})-\pi/2H(-\tilde{c},-\tilde{c})}e^{\pi/2S(v-\tilde{c},v-\tilde{c})}\left(\sum_{b\in B} e^{-\pi i b\widehat{\ }(b)+2\pi i b\widehat{\ }(v-\tilde{c})}\right)$$

$$= e^{\pi/2S(v,v)}\left(\sum_{b\in B} e^{\pi i b\widehat{\ }(b)-2\pi i c\widehat{\ }(\tilde{c})-\pi/2H(\tilde{c},\tilde{c})+\pi/2S(\tilde{c},\tilde{c})-\pi(S(v,\tilde{c})-H(v,\tilde{c}))+2\pi i b\widehat{\ }(v)}\right)$$

but $\frac{1}{2i}(H-S)(v,\tilde{c}) = \tilde{c}\widehat{\ }(v)$ by the proof of Theorem 2.13. Thus the exponent equals $-\pi i b\widehat{\ }(b)+2\pi i b\widehat{\ }(\tilde{c})-\pi i \tilde{c}\widehat{\ }(\tilde{c})+2\pi i \tilde{c}\widehat{\ }(v)+2\pi i b\widehat{\ }(v)$ which equals $-\pi i(b+\tilde{c})\widehat{\ }(b+\tilde{c})+2\pi i(b+\tilde{c})\widehat{\ }(v)$ as $b\widehat{\ }(b)$ is symmetric quadratic form on B because H is real on our $B \times B$. Thus we have the result. $\square$

§ 5.3 The Isogeny Theorem Absolutely

Let $f : X \to Y$ be an isogeny of abelian varieties where $X = V/L$ and $Y = V/M$ for lattice $M \supset L$ and f is induced by the identity. Let $A \oplus B = M$ be a decomposition of M. If $A \cap L \oplus B \cap L = L$ then we have a decomposition of

L. In this case we will say that f is compatible with the decomposition of M. Assume that this happens. Let $\mathcal{M}$ be an excellent sheaf on Y. Then $\mathcal{L} = f^*\mathcal{M}$ is an excellent sheaf on X. Thus we have canonical decomposition $A'(\mathcal{L})$ and $B'(\mathcal{L})$ of $H(\mathcal{L})$ and $A'(\mathcal{M})$ and $B'(\mathcal{M})$ of $H(\mathcal{M})$.

Lemma 5.6. *These decompositions are compatible in the sense of Section 4.4.*

Proof. We consider $f : M^{\perp}/L \to M^{\perp}/M = K(\mathcal{M})$. By assumption $M^{\perp}/L = A^{\widehat{}} \cap L \oplus B^{\widehat{}}/B \cap L$ and $M^{\perp}/M = A^{\widehat{}}/A \oplus B^{\widehat{}}/B$ where f preserves the decompositions. Clearly for $a \in A^{\widehat{}}/A \cap L$ the transformation $\sigma_A(a)$ of $\mathcal{L}$ induces the transformation $\sigma_A(f(a))$ of $\mathcal{M}$. Thus $\alpha(A'(\mathcal{L}) \cap N) \subseteq A'(\mathcal{M})$ and similarly with B. We need only see that $\mathrm{Ker}(f) = (\mathrm{Ker}(f) \cap A'(\mathcal{L})) \oplus (\mathrm{Ker}(f) \cap B'(\mathcal{L}))$ but $\mathrm{Ker}(f) = M/L = A/A \cap L \oplus B/B \cap L$. This proves the result. $\square$

Finally we have a definitive isogeny theorem.

Theorem 5.7. *In the above situation* $\rho_{\mathcal{L}}^{-1} f^* \rho_{\mathcal{M}}(\delta_b) = \sum_{\substack{b' \in B(\mathcal{L}) \\ f(b')=b}} \delta'_b$ *for all b in $B(\mathcal{M})$.*

Proof. By Theorem 4.7 we know this formula upto constant. Thus we only need to know C where $\rho_{\mathcal{L}}^{-1} f^* \rho_{\mathcal{M}}(\delta_0) = C(\sum_{f(b')=0} \delta_{b'})$. By Corollary 5.5 in the Fourier series expansion of $\delta_{b'}$ the term $e^{\pi/2 B(v,v)}$ (constant) occurs only δ_0 and the constant is 1. Thus $Ce^{\pi/2 B(v,v)}$ is the term of $f^* \rho_{\mathcal{M}}(\delta_0)$ but this is $e^{\pi/2 B(v,v)}$. So $C = 1$. $\square$

We will later see that this innocent looking theorem is responsible for a meriad of relations between theta functions.

§ 5.4 The Classical Notation

Let $X = V/L$ be an abelian variety with polarization H and $A \oplus B = L$ be a decomposition with respect to $E = \mathrm{Im}\, H$. Then we have the form $\hat{u}(v)$ on $B^{\widehat{}}$. This form has marvelous properties:

    a) $u^{\widehat{}}(v)$ is symmetric and

    b) $\mathrm{Im}\, u^{\widehat{}}(v)$ is negative definite.

The part a) was used in the proof of Corollary 5.5 and b) was used in the proof of Theorem 2.13.

We will consider the classical description of polarized abelian varieties in terms of a canonical basis $a_1, \ldots, a_g$ and $b_1, \ldots, b_g$ in A and B. We may write $\frac{1}{E(a_j, b_j)} b_j = -\sum_{1 \le k \le g} \tau_j^k a_k$. By the proof of Theorem 2.13 the a's form a complex basis of V. We want to relate the $g \times g$ matrix $\tau = (\tau_j^k)$ to the form $\hat{u}(v)$. Let $\frac{1}{E(a_j,b_j')} b_j = b_j'$. Then (b_j') is a basis on $B^\frown$.

Lemma 5.8. $(b_k')^\frown(b_j') = -\tau_j^k$.

Proof. $\hat{v}(w)$ is complex linear W and equals $E(w,v)$ when w is in A. Thus $(b_k')^\frown(b_j') = (b_k'^\frown)(\sum -\tau_j^l a_l) = -\sum \tau_j^l E(a_k, b_l') = -\tau_j^k$. $\qquad\square$

Therefore we have proven the first part of

Theorem 5.8. *The matrix τ is a complex symmetric $g \times g$ matrix and the imaginary part $\operatorname{Im}\tau$ is positive definite. Conversely, any such matrix arises from the above situation with fixed elementary divisors $\{e_j = E(a_j, b_j)\}$.*

Proof. Let $V = \mathbb{C}^g$ with unit basis $a_1, \ldots, a_g$. Let $b_j = e_j(\sum_{1 \le k \le g} \tau_j^k a_k)$ where $e_1 | \ldots | e_g$ is a successively divisible sequence of positive integers. Then as $\operatorname{Im}\tau_j^k$ is non-singular $L = \oplus \mathbb{Z} a_i \oplus \oplus \mathbb{Z} b_i$ is a lattice. We define a skew-symmetric form E on the real space of V by setting $E(a_j, a_k) = E(b_j, b_k) = 0 = E(a_l, b_k)$ if $l \ne k$ and $E(a_l, b_l) = e_l$. We want to write E as a matrix for the basis $a_1, \ldots, a_g, i a_1, \ldots, i a_g$. By the linear algebra this is

$$
{}^t\begin{bmatrix} 1 & \operatorname{Re} -\tau \\ 0 & \operatorname{Im} -\tau \end{bmatrix}^{-1}
\begin{bmatrix} 0 & 1 \\ -1 & 0 \end{bmatrix}
\begin{bmatrix} 1 & \operatorname{Re} -\tau \\ 0 & \operatorname{Im} -\tau \end{bmatrix}^{-1} .
$$

Using this the matrix is

$$
\begin{bmatrix} 0 & (\operatorname{Im} -\tau)^{-1} \\ -{}^t(\operatorname{Im} -\tau)^{-1} & \left(-{}^t(\operatorname{Im} -\tau)^{-1}{}^t(\operatorname{Re} -\tau)(\operatorname{Im} -\tau)^{-1} + {}^t(\operatorname{Im} -\tau)^{-1}(\operatorname{Re} -\tau)(\operatorname{Im} -\tau)^{-1} \right) \end{bmatrix} .
$$

As τ is symmetric this reduced to

$$
\begin{bmatrix} 0 & (\operatorname{Im} -\tau)^{-1} \\ -(\operatorname{Im} -\tau)^{-1} & 0 \end{bmatrix} .
$$

Thus this matrix is of type $(1,1)$ and the corresponding Hermitian form H has matrix $((\operatorname{Im}\tau)^{-1})$ with respect to the complex basis $a_1, \ldots, a_g$ which is positive definite. Thus we have constructed the required example. $\qquad\square$

We may now write our theta functions in terms of the matrix τ. Let b in $B(\mathcal{L})$ be given by $\sum_{1 \le j \le g} n_j b_j'$ for some integer n_j. Then

$$(*) \quad \eta_{\mathcal{L}}(\delta_b)(z) = e^{+\frac{\pi}{2}{}^t z(\operatorname{Im}\tau)^{-1}z} \left(\sum_{m \in \mathbb{Z}^g} e^{+\pi i^t(\tilde{e}m+n)\tau(\tilde{e}m+n) + 2\pi i^t(\tilde{e}m+n)z} \right)$$

where $\tilde{e}$ is the diagonal matrix with entrees $e_1, \ldots, e_g$.

It will turn out that the τ's are natural analytic parameters for the family of polarized abelian varieties. The matrices H and S are not complex analytic functions of τ. So they did not appear in the classical theory which ignored the complex differential metric geometry. We will eventually eliminate them when we do moduli.

§ 5.5 The Length of the Theta Functions

Let $A \oplus B = L$ be a decomposition of the lattice of a polarized abelian variety $X = V/L$ with polarization H. Let $\mathscr{L}$ be the excellent sheaf given by H. We have constructed a natural $H(\mathscr{L})$-isomorphism $\rho_{\mathscr{L}} : \mathbb{C}(B(\mathscr{L})) \to \Gamma(X, \mathscr{L})$. We intend to compute the length $\|\rho_{\mathscr{L}}\|$ of our theory of theta functions $\rho_{\mathscr{L}}$ as introduced at the end of Section 4.3.

We need to have an answer for this length. As we have used many times we have an isomorphism $A \otimes_{\mathbb{Z}} \mathbb{C} \xrightarrow{\approx} V$. Thus $\wedge^g A \otimes_{\mathbb{Z}} \mathbb{C} \cong \wedge^g V$ where g is the dimension of X. The free $\mathbb{Z}$-module $\wedge^g A$ has two generators $\pm k$. We may identify the dual of $\pm k$ with an invariant holomorphic g-form $\pm \omega$ on X. The needed invariant is $\sqrt{|\int_X \omega \wedge \overline{\omega}|}$. We will call this the geometric height of the decomposed polarized abelian variety (X, H) with respect to A and denote it by $h(X, H, A)$.

Theorem 5.9. $\|\rho_{\mathscr{L}}\|^2 = 2^g/(\det_L E)^{1/4} \cdot h(X, H, A)$.

Proof. We will choose a canonical basis $a_1, \ldots, a_g$ and $b_1, \ldots, b_g$ of A and B. Let $\tilde{e}$ be the diagonal matrix with coefficients $E(a_j, b_j)$. So we may assume that $a_1, \ldots, a_g$ are the unit basis for $V = \mathbb{C}^g = \{(z_i)\}$ and $b_j = -\tilde{e}_j \sum_{1 \le k \le g} \tau_j^k a_k$. The height $h(X, H, A)$ is just the square root of absolute value of $\int_X dz_1 \wedge \ldots \wedge dz_g \wedge d\bar{z}_1 \wedge \ldots \wedge d\bar{z}_g = (\pm 1) 2^g$ volume of a fundamental domain for $\mathbb{Z}^n + \tau \tilde{e} \mathbb{Z}^n$ in $\mathbb{C}^g = \pm 2^g \prod_{1 \le j \le n} \tilde{e}_j \det(\mathrm{Im}\,\tau) = \pm 2^g \det \tilde{e} \det(\mathrm{Im}\,\tau)$. Thus $h(X, H, A) = (2^g \det \tilde{e} \det(\mathrm{Im}\,\tau))^{1/2}$.

By definition $\|\rho_{\mathscr{L}}\|^2 = \int_X e^{-\pi H(v,v)} \rho_{\mathscr{L}}(\delta_0) \overline{\rho_{\mathscr{L}}(\delta_0)} dv$ where dv is the invariant measure on V normalized such that $\int_X dv = 1$. In coordinates this integral is

$$\int_X e^{\pi\,{}^t z (\mathrm{Im}\,\tau)^{-1} \bar{z}} e^{\pi\,\mathrm{Re}({}^t z (\mathrm{Im}\,\tau)^{-1} z)} \left(\sum_{m,p \in \mathbb{Z}^t} e^{\pi i [{}^t m \tilde{e} \tau \tilde{e} m - {}^t p \tilde{e} \bar{\tau} \tilde{e} p] + 2\pi i [{}^t m \tilde{e} z - {}^t p \tilde{e} z]} \right) dv \ .$$

Write $z = x + iy$ where x and y are real. Thus as τ is a symmetric matrix the integral becomes

$$\int_X e^{2\pi\,{}^t y (\mathrm{Im}\,\tau)^{-1} y} \left(\sum_{m,p \in \mathbb{Z}^g} e^{\pi i [{}^t m \tilde{e} \tau \tilde{e} m - {}^t p \tilde{e} \bar{\tau} \tilde{e} p]} e^{2\pi i [({}^t m - {}^t p) \tilde{e} x + i\,{}^t (m+p) \tilde{e} y]} \right) dv \ .$$

Consider the homomorphism $\sigma : X \to \mathbb{R}^g / \operatorname{Im} \tau \cdot \tilde{e} \mathbb{Z}^g \equiv Y$ defined by $\sigma(z) = y$. Then $\operatorname{Ker}(\sigma) = \mathbb{R}^g / \mathbb{Z}^g$ and we may compute the integral by integration over the fibers of σ. As $\int_0^1 e^{2\pi i r x} dx = 0$ if $r \neq 0$ is an integer and $\int_0^1 dx = 1$. Our integral becomes $\int_Y e^{-2\pi {}^t y (\operatorname{Im} \tau)^{-1} y} \left(\sum_{m \in \mathbb{Z}^g} e^{-2\pi {}^t m \tilde{e} (\operatorname{Im} \tau) \tilde{e} m - 4\pi {}^t m \tilde{e} y} \right) dy$ where dy is the invariant measure on Y such that $\int_Y dy = 1$. Let $y = (\operatorname{Im} \tau) \tilde{e} \varkappa$. Changing variables in our integral we may write it as

$$\int_{\mathbb{R}^g / \mathbb{Z}_g} e^{-2\pi {}^t \varkappa \tilde{e} (\operatorname{Im} \tau) \tilde{e} \varkappa} \left(\sum_{m \in \mathbb{Z}^g} e^{-2\pi {}^t m \tilde{e} (\operatorname{Im} \tau) \tilde{e} m + 4\pi {}^t m \tilde{e} (\operatorname{Im} \tau) \tilde{e} \varkappa} \right) d\varkappa$$

where $d\varkappa$ is the Euclidian volume element in $\mathbb{R}^g$. Now we may write this as

$$\int_{\mathbb{R}^g / \mathbb{Z}^g} \sum_{m \in \mathbb{Z}^g} e^{-2\pi {}^t (\varkappa + m) \tilde{e} (\operatorname{Im} \tau) \tilde{e} (\varkappa + m)} d\varkappa = \int_{\mathbb{R}^g} e^{-2\pi {}^t \varkappa \tilde{e} (\operatorname{Im} \tau) \tilde{e} \varkappa} d\varkappa$$

because τ is symmetric. By the following theorem this is $\det \left(\frac{2\pi \operatorname{Im} \tau}{\pi} \right)^{-1/2} / \det \tilde{e}$ $= 2^{g/2} \det(\operatorname{Im} \tau)^{1/2} / \det \tilde{e} = \frac{2^g}{(\det \tilde{e})^{1/2}} \frac{1}{h(X, H, A)}$. The result follows from $(\det \tilde{e})^2 = |\det_L E|$. $\qquad \square$

We need

Theorem 5.10. *If A is symmetric $g \times g$ matrix with $\operatorname{Re} A$ positive definite and b is a g vector, then $\int_{\mathbb{R}^n} e^{-({}^t x A x + 2 {}^t b x)} dx = \det(\pi A^{-1})^{1/2} e^{{}^t b A^{-1} b}$ where the square root is positive if A is real.*

Proof. By analytic continuation we may assume that A and b are real. Clearly the problem is independent of the choice of coordinates in $\mathbb{R}^g$. So we may assume that $A = 1_n$. Thus our integral is

$$\int_{\mathbb{R}^n} e^{-(\sum x_j^2 + 2 \sum b_j x_j)} \prod_{1 \leq j \leq g} dx_j = \prod_{1 \leq j \leq g} \int_{\mathbb{R}} e^{-(x^2 + 2 b_j x)} dx \; .$$

A little reflexion verfies that we need only treat the case $g = 1$. Thus we want to show that $\int_{\mathbb{R}} e^{-(a x^2 + 2 b x)} dx = (\pi/a)^{1/2} e^{b^2/a}$ when $a > 0$. By completing the square in the numerator and making a linear change of variable, one is reduced to the case where $b = 0$ and $a = 1$. This last case is done by Gauss's well-known polar coordinate trick. $\qquad \square$

Chapter 6. The Algebra of the Theta Functions

§6.1 The Addition Formula

Let $X = V/L$ be an abelian variety. We will fix a decomposition $L = A \oplus B$ with respect to an ample invertible sheaf $\mathscr{L} = \mathscr{L}(\alpha, H)$ which is excellent. In this section we will consider a special case of the isogeny Theorem 5.7.

Let n be a positive integer. Let Y be the product X^n of X with itself n-times. Then we have the sheaf $\mathscr{L}(c) = \bigotimes_{1 \leq j \leq n} \pi_j^* \mathscr{L}^{\otimes c_j}$ on Y where all c_j are positive integers. Clearly $\mathscr{L}(c)$ is ample and excellent with respect to the product decomposition $A^n \oplus B^n$ of the lattice L^n of $Y = V^n/L^n$.

Now let $d = (d_1, \ldots, d_n)$ be another such sequence. Let C (resp. D) be the diagonal $n \times n$ matrix of integers with entrees $c_1, \ldots, c_n$ (resp. $d_1, \ldots, d_n$). Let F be an $n \times n$ matrix of integers. Then F define a homomorphism $F : Y \to Y$ where $F(x_1, \ldots, x_n) = (\sum_{1 \leq j \leq n} F_i^j x_j)$.

Lemma 6.1. *$\mathscr{L}(c)$ is isomorphic to $F^* \mathscr{L}(d)$ if and only if ${}^t F \cdot D \cdot F = C$. In which case F is an isogeny.*

Proof. To check that two excellent sheaves are isomorphic it is enough to see that they have the same Hermitian form or what is the same as the same ϕ. Now $\phi_{\mathscr{L}(c)} = C \cdot \phi_{\mathscr{L}}$ and $\phi_{F_*\mathscr{L}(d)} = F^{\smallfrown} \circ \phi_{\mathscr{L}(d)} \circ F = {}^t F \cdot D \cdot F \cdot \phi_{\mathscr{L}}$. Thus we have the equivalence because $\phi_{\mathscr{L}}$ is an isogeny. The last fact follows because $\det F \neq 0$. $\qquad\square$

We will assume that the equivalent conditions of Lemma 1 are satisfied. The isogeny F satisfies the compatibility condition of Section 5.3 with respect to the decomposition $A^n \oplus B^n$ in the target and source. So we may apply Theorem 5.7 to this case. Here $B(\mathscr{L}(c)) = \bigoplus_{1 \leq j \leq n} B\left(\mathscr{L}^{\otimes c_j}\right)$ and similary with B and d. By the Künneth formula we have $\Gamma(Y, \mathscr{L}(c)) = \bigotimes_{1 \leq j \leq n} \Gamma\left(X, \mathscr{L}^{\otimes c_j}\right)$ and $\mathbb{C}[B(\mathscr{L}(c))] = \bigotimes_{1 \leq j \leq n} \mathbb{C}\left[B\left(\mathscr{L}^{\otimes c_j}\right)\right]$. For theories of theta functions we have

Lemma 6.2. $\rho_{\mathscr{L}(c)} = \bigotimes_{1 \leq j \leq n} \rho_{\mathscr{L}^{\otimes c_j}}.$

Proof. The theory of theta functions are explicit expressions given in Chap. 5. To check this equation one simply applies the definitions. $\qquad\square$

Thus we have in terms of these product decompositions.

Theorem 6.3 (addition). $\rho_{\mathscr{L}(c)}^{-1} F^* \rho_{\mathscr{L}(d)} \delta_b = \sum_{\substack{b' \in B(\mathscr{L}(c)) \\ F(b')=b}} \delta_{b'}$ *for all b in*

$B(\mathscr{L}(d))$ *where* $\delta_b = \bigotimes_{1 \le j \le n} \delta_{b'_j}$ *where* $(b'_j) = b$ *with* b'_j *in* $B(\mathscr{L}^{\otimes c_i})$.

Proof. Just apply Theorem 5.7. $\qquad\square$

Thus F^* is simply $\otimes \delta_{b_j}$ goes to $\sum_{\substack{b' = B(\mathscr{L}(c)) \\ f(b')=b}} \otimes \delta_{b'_j}$. We will give some special cases but we regard the theories of theta functions ρ_* as identifications.

Let m be a non-zero integer. We take $n = 1 = d$ and $F = m$. Thus $d = m^2$. Consequently $(m)^* \mathscr{L} \approx \mathscr{L}^{\otimes m^2}$. Our formula became the classical m-plication formula

Theorem 6.4. $\rho_{\mathscr{L}^{\otimes m^2}}^{-1} \circ m^* \circ \rho_{\mathscr{L}}(\delta_b) = \sum_{\substack{b' \in B(\mathscr{L}^{\otimes m^2}) \\ mb'=b}} \delta_{b'}$ *for all b in* $B(\mathscr{L})$.

Let (d_1, d_2) be two positive integers. Let F be the matrix $\begin{bmatrix} 1 & +d_2 \\ 1 & -d_1 \end{bmatrix}$. Then ${}^t F \cdot D \cdot F = C$ where $c = (d_1 + d_2, d_1 d_2 (d_1 + d_2))$. If η denotes $\rho_{\mathscr{L}(c)}^{-1} F^* \rho_{\mathscr{L}(d)}$ which is just F^* in coordinates we have

Theorem 6.5. *Then for b_1 in $B(\mathscr{L}^{\otimes d_1})$ and b_2 in $B(\mathscr{L}^{\otimes d_2})$*

$$\eta \left(\delta_{b_1} \otimes \delta_{b_2} \right) = \sum_{\substack{b'_1 + b'_2 = b_1 \\ d_2 b'_1 - d_1 b'_2 = b_2}} \delta_{b'_1} \otimes \delta_{b'_2}$$

where b'_1 is in $B(\mathscr{L}^{\otimes d_1 + d_2})$ and b'_2 is in $B(\mathscr{L}^{\otimes d_1 d_2 (d_1 + d_2)})$.

Now we will assume that $\mathscr{L}$ is $(d_1 + d_2)$-power of an invertible sheaf. Thus $B(\mathscr{L})$ contains all the $(d_1 + d_2)$-torsion in $B(\mathscr{L}^{\otimes d_1 d_2 (d_1 + d_2)})$. In this situation we will give a basis of $\mathbb{C}[B(\mathscr{L}^{\otimes d_1}) \times B(\mathscr{L}^{\otimes d_2})]$ and $\mathbb{C}[B(\mathscr{L}^{\otimes d_1 + d_2}) \times B(\mathscr{L}^{\otimes d_1 d_2 (d_1 + d_2)})]$ such that η has a simple diagonal form.

Let J be the group of $(d_1 + d_2)$-torsion in $B(\mathscr{L})$. For b_1 in $B(\mathscr{L}^{\otimes d_1})$ and b_2 in $B(\mathscr{L}^{\otimes d_2})$ and χ a character of J let $X(b_1, b_2, \chi) = \sum_{j \in J} \chi(j) \delta_{b_1 + j} \otimes \delta_{b_2 + d_2 j}$. As (b_1, b_2, χ) runs through $[B(\mathscr{L}^{\otimes d_1}) \times B(\mathscr{L}^{\otimes d_2})]/\{(j, d_2 j)\} \times J^*$ the $X(b_1, b_2, \chi)$ form a basis of $\mathbb{C}[B(\mathscr{L}^{\otimes d_1}) \times B(\mathscr{L}^{\otimes d_2})]$.

For b_1' in $B(\mathscr{L}^{\otimes d_1+d_2})$ let $Y(b_1',\chi) = \sum_{j\in J}\chi(j)\delta_{b_1'+j}$. Also for b_2' in $B(\mathscr{L}^{\otimes d_1 d_2(d_1+d_2)})$ let $Z(b_2',\chi) = \sum_{j\in J}\chi(j)\delta_{b_2'+j}$. Then as (b_1,χ) runs through $B(\mathscr{L}^{\otimes d_1+d_2})/J \times J^*$ then $Y(b_1',\chi)$ forms a basis of $\mathbb{C}[B(\mathscr{L}^{\otimes d_1+d_2})]$ and similarly with $\{Z(b_2',\chi)\}$.

Our addition formula gives

Theorem 6.6. $\eta(X(b_1,b_2,\chi) = Y(b_1',\chi) \otimes Z(b_2',\chi)$ if $b_1 = b_1' + b_2'$ and $b_2 = d_2 b_1' - d_1 b_2'$.

Proof. By Theorem 6.5 $\eta(X(b_1,b_2,\chi)) = \sum \chi(j)\delta_{c_1} \otimes \delta_{c_2}$ where $c_1 + c_2 = b_1 + j$ and $b_2 + d_2 j = d_2 c_1 - d_1 c_2$ for some j in J and c_1 in $B(\mathscr{L}^{\otimes d_1+d_2})$ and c_2 in $B(\mathscr{L}^{\otimes d_1 d_2(d_1+d_2)})$. Thus (c_1,c_2) runs through $(b_1' + j + j', b_2' - j')$ with j and j' in J as J is the $(d_1 + d_2)$-torsion in $B(\mathscr{L}^{\otimes d_1 d_2(d_1+d_2)})$. Thus

$$\eta(X(b_1,b_2,\chi)) = \sum_{j,j'\in J} \chi(j+j')\chi(-j')\delta_{b_1'+j+j'} \otimes \delta_{b_2'-j'}$$

$$= \left(\sum_{k\in J}\chi(k)\delta_{b_1'+k}\right)\left(\sum_{l\in J}\chi(l)\delta_{b_2'+l}\right)$$

$$= Y(b_1',\chi) \otimes Z(b_2',\chi) . \qquad \square$$

§ 6.2 Multiplication

Continuing with the notation of the last section we will assume that $d_1 = d_2 = 1$. Thus $\mathscr{L}$ is a square and we have an isomorphism $F^*(\pi_1^*\mathscr{L}\otimes\pi_2^*\mathscr{L}) \approx \pi_1^*\mathscr{L}^{\otimes 2} \otimes \pi_2^*\mathscr{L}^{\otimes 2}$ where $F(x_1,x_2) = (x_1 + x_2, x_1 - x_2)$. In this case $B(\mathscr{L}^{\otimes d_1+d_2}) = B(\mathscr{L}^{\otimes d_1 d_2(d_1+d_2)}) = B(\mathscr{L}^{\otimes 2})$ and $Z = Y$ and Theorem 6.6 gives the global effect of F^* as $\eta(X(b_1,b_2,\chi)) = Y(b_1',\chi) \otimes Y(b_2',\chi)$ where b_*' are in $B(\mathscr{L}^{\otimes 2})$ and $F(b_1',b_2') = (b_1,b_2)$.

The next idea is to consider the global effect of the inclusion i of X as $X \times 0$ in $X \times X$. As $F \circ i : X \to X^2$ is the diagonal, $(F \circ i)^* : \Gamma(X,\mathscr{L}) \otimes \Gamma(X,\mathscr{L}) \to \Gamma(X,\mathscr{L}^{\otimes 2})$ is just multiplication. On the other hand $(F \circ i)^*$ is just (evaluation of the second factor at zero)$\circ\eta$. Before we write the formula for multiplication we will make a more or less trivial generalization of these ideas.

The problem is to compute the multiplication $M_{x_1,x_2} : \Gamma(X,T_{x_1}^*\mathscr{L}) \otimes \Gamma(X,T_{x_2}^*\mathscr{L}) \to \Gamma(X,T_{x_1}^*\mathscr{L} \otimes T_{x_2}^*\mathscr{L})$ for arbitrary points x_1 and x_2 of X. As F is surjective we have points y_1 and y_2 of X such that $F(y_1,y_2) = (x_1,x_2)$. Thus we have a commutative diagram

$$\begin{array}{ccc} T_{(y_1,y_2)} : X \times X & \longrightarrow & X \times X \\ \downarrow F & & \downarrow F \\ T_{(x_1,x_2)} : X \times X & \longrightarrow & X \times X . \end{array}$$

We have a covering diagram of sheaves

$$\pi_1^*(T_{y_1}^*\mathscr{L}^{\otimes 2}) \otimes \pi_2^*(T_{y_2}^*\mathscr{L}^{\otimes 2}) \longleftarrow \pi_2^*\mathscr{L}^{\otimes 2} \otimes \pi_2^*\mathscr{L}^{\otimes 2}$$
$$\uparrow \qquad\qquad\qquad\qquad\qquad\qquad \uparrow$$
$$T_{x_1}^*\mathscr{L} \otimes T_{x_2}^*\mathscr{L} \qquad \longleftarrow \qquad \pi_1^*\mathscr{L} \otimes \pi_2^*\mathscr{L} \quad .$$

Taking global section we may consider the horizontal arrows as identification via translations and the two vertical arrows contain the same information. Now composing with i we see that the multiplication has the form

$$M_{x_1,x_2,y_1,y_2} : \Gamma(X, T_{x_1}^*\mathscr{L}) \otimes \Gamma(X, T_{x_2}^*\mathscr{L}) \longrightarrow \Gamma(X, T_{y_1}^*\mathscr{L}^{\otimes 2})$$

which can be computed by (evaluation of the second factor at $-y_2$) $\circ\ \eta$. With the above identifications we have

Theorem 6.7. $M_{x_1,x_2,y_1,y_2}\left(X(b_1,b_2,\chi)\right) = (Y(b_2',\chi)|_{-y_2})Y(b_1',\chi)$.

Let $\mathscr{P}$ be the Poincaré sheaf on $X \times X\widehat{\ }$ where $X\widehat{\ }$ is the dual abelian variety. Let $\mathscr{P}_\alpha \equiv \mathscr{P}|_{X \times \alpha}$ for some point α in $X\widehat{\ }$. Then $\mathscr{P}_\alpha$ is equivalent to zero. With this notation we have

Theorem 6.8. *Let $\mathscr{M}$ be an ample invertible sheaf on X.*
  a) For α in a non-empty subset of $X\widehat{\ }$ the multiplication

$$\Gamma\left(X, \mathscr{M}^{\otimes 2} \otimes \mathscr{P}_{\alpha+\gamma}\right) \otimes \Gamma\left(X, \mathscr{M}^{\otimes 2} \otimes \mathscr{P}_{\beta-\gamma}\right) \longrightarrow \Gamma\left(X, \mathscr{M}^{\otimes 4} \otimes \mathscr{P}_{\alpha+\beta}\right)$$

is surjective for fixed β and γ in $X\widehat{\ }$.
  b) For α in a nonempty subset of $X\widehat{\ }$ the multiplication

$$\Gamma\left(X, \mathscr{M}^{\otimes 2} \otimes \mathscr{P}_\beta\right) \otimes \Gamma\left(X, \mathscr{M}^{\otimes 2} \otimes \mathscr{P}_{\alpha+\gamma}\right) \longrightarrow \Gamma\left(X, \mathscr{M}^{\otimes 4} \otimes \mathscr{P}_{\beta+\alpha+\gamma}\right)$$

is surjective for fixed β and γ in $X\widehat{\ }$.
  c) If $n \geq 2$ and $m \geq 3$ then the multiplication

$$\Gamma\left(X, \mathscr{M}^{\otimes m} \otimes \mathscr{P}_\beta\right) \otimes \Gamma\left(X, \mathscr{M}^{\otimes n} \otimes \mathscr{P}_\gamma\right) \longrightarrow \Gamma\left(X, \mathscr{M}^{\otimes n+m} \otimes \mathscr{P}_{\beta+\gamma}\right)$$

is surjective for arbitrary β and γ in $X\widehat{\ }$.

Corollary 6.9. *The graded ring $\bigoplus_{n\geq 0} \Gamma(X, \mathscr{M}^{\otimes n})$ is generated as a $\mathbb{C}$-algebra by $\Gamma(X, \mathscr{M})$ if $\mathscr{M}$ is an m-power with $m \geq 3$.*

Proof. Without loss of generality we may assume that $\mathscr{M}$ is excellent. In the previous theory let $\mathscr{L} = \mathscr{M}^{\otimes 2}$. For a) we take x_1 and x_2 such that $2\phi_{\mathscr{M}}(x_1) = \alpha + \gamma$ and $2\phi_{\mathscr{M}}(x_2) = \beta - \alpha$. Thus the multiplication in a) is M_{x_1,x_2}. Find y_1, y_2 above. Then we want M_{x_1,x_2,y_1,y_2} to be surjective. By Theorem 6.7 if for each b_1' in $B(\mathscr{L}^{\otimes 2})$ and χ in J^*, we have some b_2' such that $Y(b_2',\chi)|_{-y_2} \neq 0$ where $F(b_1',b_2')$ is in $B(\mathscr{L})^{\otimes 2}$. Given b_1' we can take $b_2' = b_1'$. Then $Y(b_2',\chi)$ is a non-zero section of $\mathscr{L}^{\otimes 2}$. Hence its value is non-zero in nonempty open subset of y_2 in $X^\wedge$. We need to note that such a y_2 is possible. Now $2y_2 = x_1 - x_2$. Thus $4\phi_{\mathscr{A}}(y_2) = (\alpha + \gamma) - (\beta - \alpha) = 2\alpha + \gamma - \beta$. Thus as $\phi_{\mathscr{A}}$ is an isogeny y_2 varies all over X as α varies in $X^\wedge$. Thus the multiplication is surjective for general α by Theorem 6.7 as the $Y(b_1,\chi)$ span the image of multiplication. This proves a) and the proof of b) is almost the same.

For c) we will consider the case $n = 2$ and $m = 3$ (the general case follows from the same kind of argument inductively applied). By Lemma 4.6, we have

$$\Gamma\left(X, \mathscr{M}^{\otimes 5} \otimes \mathscr{P}_{\beta+\gamma}\right) = \sum_{\delta \in X^\wedge} \Gamma\left(X, \mathscr{M} \otimes \mathscr{P}_{\beta+\delta}\right) \Gamma\left(X, \mathscr{M}^{\otimes 4} \otimes \mathscr{P}_{\gamma-\delta}\right)$$

$$= \sum_{\delta} \Gamma\left(X, \mathscr{M} \otimes \mathscr{P}_{\beta+\delta}\right) \Gamma\left(X, \mathscr{M}^{\otimes 2} \otimes \mathscr{P}_{-\delta}\right)$$

$$\times \Gamma\left(X, \mathscr{M}^{\otimes 2} \otimes \mathscr{P}_\gamma\right) \quad [\delta \text{ general}]$$

by b) and the fact that linear spaces are closed and the symbols depend continuously on δ. Now this is contained in the subspace $\Gamma(X, \mathscr{M}^{\otimes 3} \otimes \mathscr{P}_\beta)\Gamma(X, \mathscr{M}^{\otimes 2} \otimes \mathscr{P}_\gamma)$. Thus c) is true. $\square$

§ 6.3 Some Bilinear Relations

Let $\mathscr{L}$ be excellent with respect to the decomposition $L = A \oplus B$. The problem is to describe the kernel $R(\mathscr{L},\mathscr{L})$ of the multiplication $m : \Gamma(X,\mathscr{L}) \otimes \Gamma(X,\mathscr{L}) \to \Gamma(X,\mathscr{L}^{\otimes 2})$.

In this section we will assume that $\mathscr{L}$ is n-power of an invertible sheaf where n is an even integer ≥ 4. By Theorem 6.8 c) we know that m is surjective. On the other hand in Theorem 6.7 we have seen that m is diagonal with respect to the correct bases in the two spaces. This means that for any χ and a given b_1' in $B(\mathscr{L}^{\otimes 2})$ there exists a b_2' in $B(\mathscr{L}^{\otimes 2})$ such that $Y(b_2',\chi)|_0 \neq 0$ where $b_1' + b_2'$ and $b_1' - b_2'$ are in $B(\mathscr{L})$. Clearly the set of these b_2' are the coset $b_1' + B(\mathscr{L})$. Therefore we have proven

Lemma 6.10. *For each χ, any $B(\mathscr{L})$ coset in $B(\mathscr{L}^{\otimes 2})$ contain an element b such that $Y(b,\chi)|_0 \neq 0$.*

Furthermore consider $R(\mathscr{L},\mathscr{L}) \cap m^{-1}(\mathbb{C}Y(b_1',\chi)) \equiv S(b_1',\chi)$. As m is diagonal, $R(\mathscr{L},\mathscr{L})$ is spanned by the $S(b_1',\chi)$. Also let b be an element of $b_1' + B(\mathscr{L})$ such that $Y(b,\chi)|_0 \neq 0$. Then by Theorem 6.7 $S(b_1',\chi)$ is spanned by $Y(b,\chi)|_0 X(c + b_1', c - b_1', \chi) - Y(c,\chi)|_0 X(b + b_1', b - b_1', \chi)$ when c runs through

$b_1' + B(\mathscr{L})$. As a the above expression is a relation whenever $Y(b,\chi) \neq 0$ we have proven

Proposition 6.11. $R(\mathscr{L},\mathscr{L})$ *is the span of the expressions*

$$Y(a,\chi)|_0 X(c+b,c-b,\chi) - Y(b,\chi)|_0 X(c+a,c-a,\chi)$$

where $a \equiv b \equiv c((B(\mathscr{L})))$ are elements of $B(\mathscr{L}^{\otimes 2})$.

Now by Theorem 6.7 we have by evaluating at zero $X(b+c,b-c,\chi)|_0 = Y(b,\chi)|_0 \cdot Y(c,\chi)|_0$ if $b \equiv c((B(\mathscr{L})))$ and b and c in $B(\mathscr{L}^{\otimes 2})$. Therefore by Lemma 6.10 and Proposition 6.11 we may conclude

Theorem 6.12. $R(\mathscr{L},\mathscr{L})$ *is the span of the expression*

$$X(a+d,a-d)|_0 X(c+b,c-b) - X(b+d,b-d)|_0 X(c+a,c-a)$$

where $a \equiv b \equiv c \equiv d((B(\mathscr{L})))$ are elements of $B(\mathscr{L}^{\otimes 2})$.

This theorem is better than the proposition because the value of the Xs are more elementary than the value of the Ys.

Next we will see that in the above situation that the bilinear relations $R(\mathscr{L},\mathscr{L})$ generate all homogeneous forms vanishing on X when we embed X in projective space via $\mathscr{L}$.

Let $\mathscr{M}$ be an ample sheaf on X. If $\mathscr{M}$ is an m-power with $m \geq 3$. By Theorem 6.9 we have a surjection

$$\mathrm{Sym}_{\mathbf{C}}[\Gamma(X,\mathscr{M})] \longrightarrow \bigoplus_{n \geq 0} \Gamma(X,\mathscr{M}^{\otimes n}) \ .$$

Let I be the kernel of this surjection. In the next section we will prove

Theorem 6.13. *a) If $m = 3$ then I is generated as an ideal by its forms of degree 2 and 3.*

b) If $m \geq 4$ then I is generated as an ideal by its forms of degree 2.

Remark. Of course the form of degree 2 in I is just the image of $R(\mathscr{M},\mathscr{M})(\subset \Gamma(X,\mathscr{M}) \otimes \Gamma(X,\mathscr{M}))$ in $\mathrm{Sym}^2(\Gamma(X,\mathscr{M}))$.

Exercise 1. Show that the above works for $T_x^*\mathscr{M}$ if we may replace evaluation at zero by evaluation at any point x of X (Hint: use translation).

§ 6.4 General Relations

Let $\mathcal{N}_1$ and $\mathcal{N}_2$ be two $\mathcal{O}_X$-modules. Then $R(\mathcal{N}_1, \mathcal{N}_2)$ is the kernel of the multiplication

$$\Gamma(X, \mathcal{N}_1) \otimes \Gamma(X, \mathcal{N}_2) \longrightarrow \Gamma(X, \mathcal{N}_1 \otimes \mathcal{N}_2) \ .$$

Let $\mathcal{N}$ be a fixed ample sheaf on our abelian variety X. Let $\mathcal{L}_1$, $\mathcal{L}_2$ etc be ample invertible sheaves on X which are equivalent to the l_1, l_2 etc-power of $\mathcal{N}$.

Theorem 6.14. *If $l_3 \geq 2$ and either $l_1 \geq 3$ and $l_2 \geq 4$ or $l_1 \geq 2$ and $l_2 \geq 5$, then*

$$R(\mathcal{L}_1, \mathcal{L}_2 \otimes \mathcal{L}_3) = R(\mathcal{L}_1, \mathcal{L}_2) \Gamma(X, \mathcal{L}_3) \ .$$

This will prove Theorem 6.13. For a) note that $R(\mathcal{M}, \mathcal{M})$ and $R(\mathcal{M}, \mathcal{M}^{\otimes 2})$ essentially generate I and have degree 2 and 3. For b) I is generated by $R(\mathcal{M}, \mathcal{M})$ which has degree 2.

We will prove Theorem 6.14 by a gambit similar to that used to show multiplication is surjective. Explicitly we will use

Proposition 6.15. *If $l_1 + l_2 \geq 5$ and $l_1, l_2 \geq 2$, then*

$$R(\mathcal{L}_1, \mathcal{L}_2 \otimes \mathcal{L}_3) = \sum_{\alpha \in X\hat{}} R(\mathcal{L}_1, \mathcal{L}_2 \otimes \mathcal{P}_\alpha) \Gamma(X, \mathcal{L}_3 \otimes \mathcal{P}_{-\alpha}) \ .$$

First we will show that this implies Theorem 6.14. Write

$$\mathcal{L}_2 = \mathcal{L}_4 \otimes \mathcal{L}_5 \quad \text{where } l_5 = 2 \ .$$

By Proposition 6.15

$$R(\mathcal{L}_1, \mathcal{L}_2 \otimes \mathcal{L}_3) = \sum R(\mathcal{L}_1, \mathcal{L}_4 \otimes \mathcal{P}_\alpha) \Gamma(X, \mathcal{L}_5 \otimes \mathcal{L}_3 \otimes \mathcal{P}_{-\alpha})$$

but by Theorem 6.8

$$\Gamma(X, \mathcal{L}_5 \otimes \mathcal{L}_3 \otimes \mathcal{P}_{-\alpha}) = \Gamma(X, \mathcal{L}_5 \otimes \mathcal{P}_{-\alpha}) \Gamma(X, \mathcal{L}_3)$$

for general α. Thus

$$R(\mathcal{L}_1, \mathcal{L}_2 \otimes \mathcal{L}_3) = \left(\sum_{\alpha \text{ general}} R(\mathcal{L}_1, \mathcal{L}_4 \otimes \mathcal{P}_\alpha) \Gamma(X, \mathcal{L}_5 \otimes \mathcal{P}_{-\alpha}) \right) \Gamma(X, \mathcal{L}_3)$$

$$\subseteq R(\mathcal{L}_1, \mathcal{L}_2) \Gamma(X, \mathcal{L}_3)$$

the reverse inclusion is obvious. Note that we implicitly used that $R(\mathscr{L}_1, \mathscr{L}_4 \otimes \mathscr{P}_{-\alpha})$ is a vector bundle on $X\hat{}$, which follows from the exact sequence of vector bundles

$$(*) \quad 0 \to R(\mathscr{L}_1, \mathscr{L}_4 \otimes \mathscr{P}_{-\alpha}) \to \Gamma(\mathscr{L}_1) \underset{\mathbf{C}}{\bigotimes} \Gamma(\mathscr{L}_4 \otimes \mathscr{P}_{-\alpha})$$
$$\to \Gamma(\mathscr{L}_1 \otimes \mathscr{L}_4 \otimes \mathscr{P}_{-\alpha}) \to 0$$

(Theorem 6.8). Thus it remains to check Proposition 6.15. For that look at the commutative exact diagram

$$\begin{array}{ccccccccc}
0 \to & R(\mathscr{L}_1, \mathscr{L}_2 \otimes \mathscr{L}_3) & \longrightarrow & \Gamma(\mathscr{L}_1) \otimes \Gamma(\mathscr{L}_2 \otimes \mathscr{L}_3) & \longrightarrow & \Gamma(\mathscr{L}_1 \otimes \mathscr{L}_2 \otimes \mathscr{L}_3) & \longrightarrow & 0 \\
& \uparrow & & \uparrow & & \uparrow & & \\
0 \to & \begin{array}{c} R(\mathscr{L}_1, \mathscr{L}_2 \otimes \mathscr{P}_\alpha) \otimes \\ \Gamma(\mathscr{L}_3 \otimes \mathscr{P}_{-\alpha}) \end{array} & \to & \begin{array}{c} \Gamma(\mathscr{L}_1) \otimes \Gamma(\mathscr{L}_2 \otimes \mathscr{P}_\alpha) \otimes \\ \Gamma(X, \mathscr{L}_3 \otimes \mathscr{P}_{-\alpha}) \end{array} & \to & \begin{array}{c} \Gamma(\mathscr{L}_1 \otimes \mathscr{L}_2 \otimes \mathscr{P}_\alpha) \otimes \\ \Gamma(\mathscr{L}_3 \otimes \mathscr{P}_{-\alpha}) \end{array} & \to & 0
\end{array}$$

We need to show that any linear functional λ on $\Gamma(X, \mathscr{L}_1) \otimes \Gamma(X, \mathscr{L}_2 \otimes \mathscr{L}_3)$ which induces zero on $R(\mathscr{L}_1, \mathscr{L}_2 \otimes \mathscr{P}_\alpha) \otimes \Gamma(X, \mathscr{L}_3 \otimes \mathscr{P}_{-\alpha})$ for all α in $X\hat{}$ is induced by a linear functional μ on $\Gamma(X, \mathscr{L}_1 \otimes \mathscr{L}_2 \otimes \mathscr{L}_3)$. By assumption λ induces a linear functional on $\Gamma(X, \mathscr{L}_1) \otimes \Gamma(X, \mathscr{L}_2 \otimes \mathscr{P}_\alpha) \otimes \Gamma(X, \mathscr{L}_3 \otimes \mathscr{P}_{-\alpha})$ which comes from one say μ_α on $\Gamma(X, \mathscr{L}_1 \otimes \mathscr{L}_2 \otimes \mathscr{P}_\alpha) \otimes \Gamma(X, \mathscr{L}_3 \otimes \mathscr{P}_{-\alpha})$. By Lemma 4.6 μ if it exists is uniquely determined by the family $\{\mu_\alpha\}$. On the other hand, $\{\mu_\alpha\}$ clearly depends regularly on α in $X\hat{}$. If we could show that such a family is induced by such a μ we would be done (just apply Lemma 4.6 to the second arrow to check that these μ induces λ).

Next we write our problem in terms of sheaf theory on $X\hat{}$. For any ample sheaf $\mathscr{L}$ on X let $W^{\pm}(\mathscr{L}) = \pi_{X\hat{}*}(\pi_X^* \mathscr{L} \otimes \mathscr{P}^{\otimes \pm 1})$ be a locally free sheaf on $X\hat{}$ whose fibers at α are $\Gamma(X, \mathscr{L} \otimes \mathscr{P}_{\pm\alpha})$. We have multiplication

$$W^+(\mathscr{L}_1) \otimes W^-(\mathscr{L}_2) \longrightarrow \pi_{X\hat{}*}(\pi_X^*(\mathscr{L}_1 \otimes \mathscr{L}_2)) = \Gamma(X, \mathscr{L}_1 \otimes \mathscr{L}_2) \otimes \mathcal{O}_{X\hat{}}$$

for two ample sheaves $\mathscr{L}_1$ and $\mathscr{L}_2$ on X. Thus we have the dual homomorphism

$$M : \Gamma(X, \mathscr{L}_1 \otimes \mathscr{L}_2)\hat{} \longrightarrow \Gamma\left(X\hat{}, (W^+(\mathscr{L}_1) \otimes W^-(\mathscr{L}_2))\hat{}\right) .$$

Our problem is just to show

Proposition 6.16. *M is an isomorphism.*

Proof. We have seen that Lemma 4.6 is equivalent to the injectivity of M. To show that M is surjective we will show that the two spaces have the same dimension.

Now $\mathcal{K} \equiv (W^+(\mathcal{L}_1) \otimes W^-(\mathcal{L}_2))\widehat{} = (W^+(\mathcal{L}_1))\widehat{} \otimes (W^-(\mathcal{L}_2))\widehat{}$ has fibers $(\Gamma(\mathcal{L}_1 \otimes \mathcal{P}_\alpha))\widehat{} \otimes (\Gamma(\mathcal{L}_2 \otimes \mathcal{P}_{-\alpha}))\widehat{} \cong H^g(\mathcal{L}_1^{\otimes -1} \otimes \mathcal{P}_{-\alpha}) \otimes H^g(\mathcal{L}_2^{\otimes -1} \otimes \mathcal{P}_\alpha)$ by Serre duality where $g = \dim X$. Thus if $\mathcal{F} = \pi_1^*(\mathcal{L}_1^{\otimes -1} \otimes \pi_{1,3}^* \mathcal{P}^{\otimes -1}) \otimes \pi_2^*(\mathcal{L}_2^{\otimes -1} \otimes \pi_{2,3}^* \mathcal{P})$ is a sheaf on $X \times X \times X\widehat{}$, we have an isomorphism $R^{2g}\pi_{3*}\mathcal{F} \cong \mathcal{K}$. As $\mathcal{L}_1$ and $\mathcal{L}_2$ are ample this is the only non-zero direct image of $\mathcal{F}$ via π_3. Thus by a degenerate Leray spectral sequence we have isomorphism $H^{2g+i}(\mathcal{F}) \cong H^i(\mathcal{K})$. Hence we need only compute the cohomology of an invertible sheaf on the abelian variety $X \times X \times X\widehat{}$.

One method to finish is to show $|\chi(\mathcal{F})| = |\chi(\mathcal{L}_1 \otimes \mathcal{L}_2)|$. Hence $\mathcal{F}$ is non-degenerate and $H^{2g}(\mathcal{F}) \cong H^0(\mathcal{K})$ is its non-zero cohomology group and has dimension $= \dim \Gamma(\mathcal{L}_1 \otimes \mathcal{L}_2)$. As this calculation is messy we will use a more dramatic way to compute the cohomology of $\mathcal{F}$. The trick is

Claim. If Δ is the diagonal in $X \times X$, $R^g\pi_{12*}\mathcal{F} \approx \pi_1^*\mathcal{L}_1^{\otimes -1} \otimes \pi_2^*\mathcal{L}_2^{\otimes -1}|_\Delta \approx (\mathcal{L}_1 \otimes \mathcal{L}_2)^{\otimes -1}$ is the only non-zero higher direct image of $\mathcal{F}$ via π_{12}.

Thus if we prove the claim then a degenerate Leray spectral sequence gives an isomorphism $H^{g+j}(\mathcal{F}) \approx H^j((\mathcal{L}_1 \otimes \mathcal{L}_2)^{\otimes -1})$ which equals $\Gamma((\mathcal{L}_1 \otimes \mathcal{L}_2))\widehat{}$ if $j = g$. Therefore $\Gamma(\mathcal{K}) \approx H^{2g}(\mathcal{F}) \approx H^g((\mathcal{L}_1 \otimes \mathcal{L}_2)^{\otimes -1}) \cong \Gamma((\mathcal{L}_1 \otimes \mathcal{L}_2))\widehat{}$ and we will be done.

To prove the claim write $\mathcal{F}$ as $\pi_{12}^*(\pi_1^*\mathcal{L}_1^{\otimes -1} \otimes \pi_2^*\mathcal{L}_2^{\otimes -1}) \otimes (\pi_3 + \pi_2, \pi_3)^*\mathcal{P}$. Thus

$$R^j\pi_{12*}\mathcal{F} \approx (\pi_1^*\mathcal{L}_1^{\otimes -1} \otimes \pi_2^*\mathcal{L}_2^{\otimes -1}) \otimes R^j\pi_{12*}(-\pi_1 + \pi_2, \pi_3)^*\mathcal{P}$$
$$\approx (\pi_1^*\mathcal{L}_1^{\otimes -1} \otimes \pi_2^*\mathcal{L}_2^{\otimes -1}) \otimes (-\pi_1 + \pi_2)^*(R^j\pi_{1*}\mathcal{P})$$

but by Theorem 3.15 $R^j\pi_{1*}\mathcal{P} \approx \mathcal{O}_0$ if $j = g$ and is otherwise zero and $(-\pi_1 + \pi_2)(0) = \Delta$. So the claim follows by the Künneth formula for higher direct images.
$\qquad\square$

Chapter 7. Moduli Spaces

§ 7.1 Complex Structures on a Symplectic Space

Let W be a real vector space of dimension $2g$ with a non-degenerate skew-symmetric form E. Recall the symplectic group $\mathrm{Sym}_{\mathbb{R}}(E)$ consists of all $\mathbb{R}$-linear transformations A of E such that $E(Ax, Ay) = E(x, y)$ for all x and y in W.

We want to consider complex structures on W. Such a complex structure (W, J) is determined by a real linear isomorphism $J : W \to W$ such that $J^2 = -1$ which gives the effect of multiplication by i in the complex structure. The symplectic group of transformations of W preserving E operates on such complex structures. For A in $\mathrm{Sym}_{\mathbb{R}}(E)$ and complex structure J on W

$$A * J = AJA^{-1} .$$

Thus the isomorphism $A : (W, J) \to (W, A * J)$ given by A is complex linear for the two different complex structures. This is just the equation $A(J(W)) = (AJA^{-1})(A(W))$.

A complex structure J on W is reasonable if the following two conditions a) and b) hold.

a) $E(Jw_1, Jw_2) = E(w_1, w_2)$ for all w_1 and w_2 in W.

By Proposition 1.2 there is a unique Hermitian form H_J on (W, J) with imaginary part E. By Corollary 1.4, H_J is non-degenerate. The other condition is

b) H_J is positive definite (equivalently, $E(J(x), x) > 0$ for all x in $W - \{0\}$).

We shall be very interested in the properties of the space $\mathrm{Reas}(W)$ of all reasonable complex structures on W. We will first see that it is a homogeneous space under $\mathrm{Sym}_{\mathbb{R}}(W)$.

Theorem 7.1. *The space* $\mathrm{Reas}(W)$ *is invariant under the action of* $\mathrm{Sym}_{\mathbb{R}}(W)$. *Furthermore* $\mathrm{Sym}_{\mathbb{R}}(W)$ *acts transitively on* $\mathrm{Reas}(W)$ *and the stabilizer of a reasonable complex structure* J *on* W *is the unitary group of the Hermitian form* H_J *on* (W, J).

Proof. The first statement is fairly routine. For A in $\mathrm{Sym}_{\mathbb{R}}(W)$ and J in $\mathrm{Reas}(W)$ we want to show that $A * J$ is in $\mathrm{Reas}(W)$. Note that

$$E(AJA^{-1}w_1, AJA^{-1}w_2) = E(JA^{-1}w_1, JA^{-1}w_2)$$
$$= E(A^{-1}w_1, A^{-1}w_2) = E(w_1, w_2) \ .$$

Next we will prove that $A : (W, J) \to (W, A * J)$ is unitary for the matrices H_J and H_{A*J}. in other words,

$$(*) \qquad\qquad H_{A*J}(Az, Aw) = H_J(z, w) \ .$$

Recall that $H_J(z, w) = E(Jz, w) + iE(z, w)$. Thus

$$H_{A*J}(Az, Aw) = E(AJA^{-1}Az, Aw) + iE(Az, Aw)$$
$$= E(Jz, Aw) + iE(z, w) = H_J(z, w) \ .$$

Thus the first statement is true.

For the transitivity we will establish a natural bijective correspondence between {symplectic bases of W} and {unitary bases in H_J with J in $\mathrm{Reas}(W)$}. As $\mathrm{Sym}_{\mathbb{R}}(W)$ acts transitively on {symplectic bases of W}, this will show that it acts transitively on J. Also it will show the second statement because if A is in the stabilizer of J, A is unitary for H_J on (W, J) by the first paragraph. Conversely unitary transformations correspond to a change of the unitary basis. So by the bijectivity any unitary transformation comes from a symplectic transformation.

Let $a_1, \ldots, a_g$, $b_1, \ldots, b_g$ be a symplectic basis of W. Then $E(a_k, a_j) = 0 = E(b_k, b_j)$ for all j and k, $E(a_k, b_j) = \delta_j^k$ for all k and j. We have an associated complex structure J such that $J(a_i) = -b_i$ and $J(b_j) = +a_j$ for all j. The associated Hermitian form H_J is

$$H_J\left(\sum_j \lambda_j a_j + \mu_j b_j, \sum_j \gamma_j a_j + \delta_j b_j\right) = +\sum_j \lambda_j \gamma_j + \mu_j \delta_j + i\sum_j \lambda_j \delta_j - \mu_j \gamma_j \ .$$

Thus H_J is positive definite; i.e. J is in $\mathrm{Reas}(W)$. Furthermore $a_1, \ldots, a_g$ are a unitary basis with respect to H_J. Conversely if $a_1, \ldots, a_g$ are a unitary basis for H_J for some reasonable complex structure J, then $a_1, \ldots, a_g, -J(a_1), \ldots, -J(a_g)$ is a symplectic basis for W. Clearly this bijection is natural with respect to the $\mathrm{Sym}_{\mathbb{R}}(W)$ action. $\qquad\qquad \square$

We have actually proven more. Consider the space $B = \{(w, J)$ where w is in W and J is in $\mathrm{Reas}(W)\}$. Then projection $\pi : B \to \mathrm{Reas}(W)$ is a complex vector bundle with respect to the complex structure (W, J) on the fiber over J. In fact we have a Hermitian structure on this bundle given by H_J on the fiber over J. We have proven

Corollary 7.2. *B is a homogeneous Hermitian bundle over* Reas(W) *with respect to the group* Sym$_{\mathbb{R}}(W)$.

In turns out that Reas(W) has the structure of a Hermitian symmetric space. In other words Reas(W) has a complex structure, Hermitian metric (all invariant under Sym$_{\mathbb{R}}(W)$) and an holomorphic isometry φ_x fixing any given point x which reverses all directions at x.

To define the complex structure we need to know the tangent space to Reas(W) at any given point J. A first order deformation of J is given by $J + \epsilon V$ where V is an endomorphism W such that

a') $(J + \epsilon V)^2 = -1((\epsilon^2))$; i.e., $JV + VJ = 0 \Leftrightarrow JV = -VJ \Leftrightarrow V$ is an anti-complex endomorphism of (W, J), and

b') $E(J+\epsilon V, J+V) = E((\epsilon^2))$; i.e., $E(V, J) + E(J, V) = 0$, or, equivalently by a') and b) $E(V, 1) + E(1, V) = 0$.

A little calculation shows that $V \mapsto -JV$ gives a complex structure on this tangent space. By a) the tangent space is a real subspace of Hom$_{\mathbb{R}}(W, W)$.

Lemma 7.3. *If V is a vector in the tangent space of* Reas(W), *let* $\|V\|^2 = \mathrm{Tr}(\widetilde{V}^2)$ *where* $\widetilde{V} : W \to W$ *is the corresponding endomorphism. The* $\| \ \|$ *is Hermitian metric on* Reas(W) *which is invariant under the action* Sym$_{\mathbb{R}}(W)$.

Proof. The invariance is easy. For A in Sym$_{\mathbb{R}}(W)$ the action on Reas(W) is given by $J \to AJA^{-1}$. Hence the differential action is $V \to AVA^{-1}$. Now $\widetilde{AVA^{-1}} = A \cdot \widetilde{V} \cdot A^{-1}$. So $\mathrm{Tr}\left((\widetilde{AVA^{-1}})^2\right) = \mathrm{Tr}(A\widetilde{V}^2 A^{-1}) = \mathrm{Tr}(\widetilde{V}^2)$. This shows invariance.

Let J be a fixed point of Reas(W). Let $a_1, \ldots, a_g$ be a unitary basis of (W, J) with respect to the form H_J. As we have seen in the last proof, $a_1, \ldots, a_g, -J(a_1), \ldots, -J(a_g)$ is a symplectic basis of W. Let V be a tangent at J. We want to determine the matrix of $\widetilde{V}$ in this basis.

Claim. The space of these matrices is the space of matrices of the form $\begin{bmatrix} K & L \\ L & -K \end{bmatrix}$ where K and L are real symmetric $g \times g$ matrices.

Proof of Claim. The condition on such real matrices $\begin{bmatrix} a & c \\ b & d \end{bmatrix}$ as a')

$$\begin{bmatrix} 0 & 1 \\ -1 & 0 \end{bmatrix}\begin{bmatrix} a & c \\ b & d \end{bmatrix} = -\begin{bmatrix} a & c \\ b & d \end{bmatrix}\begin{bmatrix} 0 & 1 \\ -1 & 0 \end{bmatrix} \text{ and } {}^t\begin{bmatrix} a & c \\ b & d \end{bmatrix}\begin{bmatrix} 0 & 1 \\ -1 & 0 \end{bmatrix}$$

$$= -\begin{bmatrix} 0 & 1 \\ -1 & 0 \end{bmatrix}\begin{bmatrix} a & c \\ b & d \end{bmatrix} .$$

The rest of the proof of this claim is obvious.

Now

$$\mathrm{Tr}\begin{bmatrix} K & L \\ L & -K \end{bmatrix}^2 = \mathrm{Tr}\begin{bmatrix} K^2 + L^2 & KL - LK \\ LK - KL & K^2 + L^2 \end{bmatrix}$$

$$= 2\,\mathrm{Tr}(K^2 + L^2) = 2\sum_{m,n}(k_n^m)^2 + (l_n^m)^2$$

as K and L are symmetric. Hence $\| \ \|$ is positive definite.

It remains to show that $\| \ \|$ is Hermitian. Now

$$i\begin{bmatrix} K & L \\ L & -K \end{bmatrix} = -\begin{bmatrix} 0 & 1 \\ -1 & 0 \end{bmatrix}\begin{bmatrix} K & L \\ L & -K \end{bmatrix} = \begin{bmatrix} -L & K \\ K & L \end{bmatrix}.$$

Thus we can think of K as the real part and L as the imaginary part of $\begin{bmatrix} K & L \\ L & -K \end{bmatrix}$. So the matrix is Hermitian. $\qquad\square$

In the next section we will show that the almost complex structure on $\mathrm{Reas}(W)$ is a complex structure by introducing coordinates.

§ 7.2 Siegel Upper-half Space

Let g be a positive integer. The Siegel space S_g is the complex manifold of all symmetric $g \times g$ complex matrices τ with positive definite imaginary part. Thus $\dim S_g = (g+1)g/2$.

Let $a_1,\ldots,a_g,\ b_1,\ldots,b_g$ be a symplectic basis of our real vector space W. Thus $W = \oplus \mathbb{R}a_j \oplus \oplus \mathbb{R}b_j$ with the standard symplectic form E.

Theorem 7.4. *We have a bijection $R : \mathrm{Reas}(W) \xrightarrow{\approx} S_g$ defined by the above choice.*

Proof. For J in $\mathrm{Reas}(W)$$((W,J)/L, H_J)$ is a principally polarized abelian variety where $L = \oplus \mathbb{Z}a_i \oplus \oplus \mathbb{R}b_j$ and $(a_1,\ldots,a_g,b_1,\ldots,b_g)$ are a canonical basis of L with respect to the skew-symmetric form $E = \mathrm{Im}\,H_J$. Thus we may apply the procedure of the Theorem 5.8 to produce the point $R(J)$ of S_g. Let (x_j^k) and (y_j^k) be the $g \times g$ real matrices such that $b_j = -\sum(x_j^k + y_j^k J)a_k$. Thus $R(J) = (x_j^k + iy_j^k) \equiv \tau$. In that theorem we have seen that $R(J)$ is in S_g and all of S_g occurs this way uniquely. $\qquad\square$

From the point of view of Siegel space the complex vector space (W,J) is isomorphic to the constant complex space $\mathbb{C}^g = \oplus \mathbb{C}a_j$ but the Hermitian structure is changing. Thus we have a holomorphic trivialization

$$\mathbb{C}^g \times S_g \xrightarrow{\approx} R^{-1}B$$
$$\pi \searrow \quad \downarrow \pi'$$
$$S_g \quad S_g$$

but the Hermitian matrix in the fiber is given by $\sum_{k,j} z_k \left((\operatorname{Im}\tau)^{-1}\right)_j^k \bar{z}_j$ by Section 5.4.

We will next work out how $\operatorname{Sym}_\mathbb{R}(W)$ acts on the bundle $\mathbb{C}^g \times S_g$ (in particular how it acts on S_g).

In terms of our basis $a_1,\ldots,a_g, b_1,\ldots,b_g$ of W an element of $\operatorname{Sym}_\mathbb{R}(W)$ is a $2g \times 2g$ matrix $\begin{bmatrix} \alpha & \beta \\ \gamma & \delta \end{bmatrix}$ where α, β, γ, δ are $g \times g$ matrices such that

$$^t\begin{bmatrix} \alpha & \beta \\ \gamma & \delta \end{bmatrix}\begin{bmatrix} 0 & 1 \\ -1 & 0 \end{bmatrix}\begin{bmatrix} \alpha & \beta \\ \gamma & \delta \end{bmatrix} = \begin{bmatrix} 0 & 1 \\ -1 & 0 \end{bmatrix}.$$

Lemma 7.5. *The action of* $\operatorname{Sym}_\mathbb{R}(W)$ *on* $C^g \times S_g$ *is given by*

$$\begin{bmatrix} a & b \\ c & d \end{bmatrix} * (z,\tau) = (z',\tau') = \left(^t(c\tau + d)^{-1}z, (a\tau + b)(c\tau + d)^{-1}\right).$$

Proof. Let $A = \begin{bmatrix} a & b \\ c & d \end{bmatrix}$. Let J be the complex structure $R^{-1}(\tau)$ on W. Then we have a commutative diagram

$$\oplus \mathbb{R}a_1 \oplus \oplus \mathbb{R}b_j = (w, J) \xrightarrow{(1,-\tau)} \mathbb{C}^g$$
$$\downarrow A \qquad\qquad \downarrow X$$
$$\oplus \mathbb{R}a_i \oplus \oplus \mathbb{R}b_j = (W, A*J) \xrightarrow{(1,-\tau')} \mathbb{C}^g$$

where the complex linear isomorphism X is given a complex $g \times g$ matrix of the same name. Therefore $X(1,-\tau) = (1,-\tau')\begin{bmatrix} a & b \\ c & d \end{bmatrix}$ or rather $(X, -X\cdot\tau) = (a - \tau'c, b - \tau'd)$. So $X = a - \tau'c$ and $\tau = -(a - \tau'c)^{-1}(b - \tau'd)$. We want to reverse the roles of τ and τ'.

The inverse of A is

$$\begin{bmatrix} 0 & -1 \\ 1 & 0 \end{bmatrix} {}^t\begin{bmatrix} a & b \\ c & d \end{bmatrix}\begin{bmatrix} 0 & 1 \\ -1 & 0 \end{bmatrix} = \begin{bmatrix} {}^td & -{}^tb \\ -{}^tc & {}^ta \end{bmatrix}.$$

The reversed equations are

$$X^{-1} = {}^td + \tau\,{}^tc \quad \text{and} \quad \tau' = +({}^td + \tau\,{}^tc)^{-1}({}^tb + \tau\,{}^ta).$$

So

$$X = ({}^td + \tau\,{}^tc)^{-1} = {}^t(d + c\,{}^t\tau)^{-1} = {}^t((d + c\tau)^{-1})$$

and

$$\tau' = {}^{t}(\tau') = (b + a^{t}\tau)(d + c^{t}\tau)^{-1} = (b + a\tau)(d + c\tau)^{-1} \ . \qquad \square$$

We next have

Corollary 7.6. *a)* $\mathrm{Sym}_{\mathbb{R}}(W)$ *acts by holomorphic transformation on* S_g.
b) $R : \mathrm{Reas}(W) \xrightarrow{\approx} S_g$ *is biholomorphic.*

Proof. a) follows directly from $\tau \rightarrow (a\tau + b)(c\tau + d^{-1})$ being holomorphic in τ.
As we are dealing with $\mathrm{Sym}_{\mathbb{R}}(W)$ homogeneous space by Theorem 7.1, for b)
we need only check that R gives a complex isomorphism on tangent space at
one point say

$$R^{-1}(iI) = \begin{bmatrix} 0 & I \\ -I & 0 \end{bmatrix} \equiv J \ .$$

First we will compute $T_{(W,J)}(V)$ for any tangent vector V of $\mathrm{Reas}(W)$ at
any complex structure J. Let ϵ be a variable with $\epsilon^2 = 0$.
Then by definition

$$R(J + \epsilon V) = R(J) + \epsilon T_{(W,J)}(V) \ .$$

Let

$$V(a_j) = \sum_k u_j^k a_k + \sum_k w_j^k J a_k \quad \text{for } 1 \le j \le g \ .$$

Then

$$R(J + \epsilon V) = (\tau_j^k) + \epsilon(\sigma_j^k)$$

where

$$(\tau_j^k) = R(J) \quad \text{and} \quad (\sigma_j^k) = T_{(W,J)}(V) \ .$$

Let

$$(\tau_j^k) = (x_j^k) + i(y_j^k) \quad \text{and} \quad (\sigma_j^k) = (r_j^k) + i(s_j^k)$$

be the real and imaginary parts. By definition

$$b_k = -\sum_j \left[(x_k^j + \epsilon r_k^j) + (y_k^j + \epsilon s_k^j)(J + \epsilon V) \right] a_j$$

$$= -\left[\sum_j (x_k^j + y_k^j J) + \epsilon(r_k^j + s_k^j V + s_k^j V + s_k^j J) \right] a_j \ .$$

Thus

$$\sum_j (r_k^j + s_k^j V + s_k^j J) a_j = 0 \quad \text{for all } 1 \le j \le g \ ,$$

or, rather

$$\sum_j (r_k^j + s_k^j J) a_j = -\sum_j y_k^j V a_j = -\sum_j y_k^j \left(\sum_l u_j^l a_k + \sum_l w_j^l J a_k \right) .$$

In terms of matrices this gives

$$R = -Y \cdot U \quad \text{and} \quad S = -Y \cdot W .$$

Thus in the particular case

$$(Y = 1) T_{(W,J)}(V) = -U - iW$$

where

$$V(a_j) = \sum_k u_j^k a_k + \sum_k w_j^k b_k .$$

Thus $T_{(W,J)}$ is injective. We need to check that $T_{(W,J)}$ is surjective and complex linear; i.e., $T_{(W,J)}(iV) = +W - iU$. In the proof of Lemma 7.3 we have seen that in terms of the basis $a_1,\ldots,a_g,b_2,\ldots,b_g$ it is given by $\begin{bmatrix} U & +W \\ +W & -U \end{bmatrix}$ and U and W are symmetric real matrices. Thus $T_{(W,J)}$ is surjective. Furthermore in that lemma we have seen that

$$i \begin{bmatrix} U & +W \\ +W & -U \end{bmatrix} = \begin{bmatrix} -W & U \\ U & W \end{bmatrix} .$$

Thus $T_{(W,J)}(iV) = -W + iU$ and hence $T_{(W,J)}$ is complex linear. $\square$

Thus we have a $\mathrm{Sym}_{\mathbb{R}}(2g)$ invariant Hermitian metric of S_g corresponding $\| \ \|$ on $\mathrm{Reas}(W)$. The homogeneous space S_g is actually a symmetric space. The symmetry about

$$iI \text{ is } \tau \to -\tau^{-1} = (A\tau + B)(C\tau + D)^{-1}$$

where

$$\begin{bmatrix} A & B \\ C & D \end{bmatrix} = \begin{bmatrix} 0 & 1 \\ -1 & 0 \end{bmatrix} .$$

Exercise 1. If $\tau = x + iy$ is in S_g, then $R^{-1}(\tau)$ is the endomorphism given by

$$\begin{bmatrix} -xy^{-1} & -(xy^{-1}x + y) \\ y^{-1} & y^{-1}x \end{bmatrix} ,$$

in the basis $a_1,\ldots,a_b, b_1,\ldots,b_g$.

§ 7.3 Families of Abelian Varieties and Moduli Spaces

Let L be a lattice in W such that the form E has integral values on $L \times L$. For each complex structure J in $\mathrm{Reas}(W)$ we have the polarized abelian variety $S(J) = ((W, J)/L, H_J)$. We can continuously construct a family of the varieties as follows. Consider $\pi' : B/L \to \mathrm{Reas}(W)$ where B/L is the bundle B modulo L. Explicitly $B/L = \{(w, J) \mid w \in W\} / / \{(w_1, J) \sim (w_2, J) \text{ if } w_1 - w_2 \in L\}$. In the last section we have seen that $\pi : B \to \mathrm{Reas}(W)$ is a trivial holomorphic bundle. So B/L is a complex manifold and π' is holomorphic, smooth and proper mapping and the polarization H_J depends continuously on the fibers. This is the basic family of polarized abelian varieties.

The basic moduli spaces are $M(W, L) \equiv \mathrm{Reas}(W)$ modulo the equivalence relation $J_1 \sim J_2$ if $S(J_1)$ and $S(J_2)$ are isomorphic polarized abelian varieties.

Lemma 7.7. *We have a natural bijection*

$$M(W, L) = \mathrm{Sym}_{\mathbf{Z}}(L) \setminus \mathrm{Reas}(W)$$

and $\mathrm{Sym}_{\mathbf{Z}}(L)$ *acts properly and discontinuously on* $\mathrm{Reas}(W)$.

Proof. For the first statement let A be an element of $\mathrm{Sym}_{\mathbf{Z}}(L)$ and J be a point in $\mathrm{Reas}(W)$. Then $A : (W, J) \to (W, A * J)$ takes L in L and H_J into H_{A*J}. Conversely if $B : X(J) \xrightarrow{\approx} X(J')$, then B is given by an isomorphism $A : W \to W$ which takes L to L and H_J to $H_{J'}$. Thus A preserves $E = \mathrm{Im}\, H_J = \mathrm{Im}\, H_{J'}$. Hence A is in $\mathrm{Sym}_{\mathbf{Z}}(L)$ and $B = \widetilde{A}$.

The second statement means that a) for all J in $\mathrm{Reas}(W)$ the stabilizer G_J of J in $\mathrm{Sym}_{\mathbf{Z}}(L)$ is finite and there is a G_J-invariant open neighborhood U_J of J such that for all A in $\mathrm{Sym}_{\mathbf{Z}}(L)$, $A \cdot U_J \cap U_J \neq \phi \Leftrightarrow A \in G_J$ and b) for J and J' in $\mathrm{Reas}(W)$ such that $\mathrm{Sym}_{\mathbf{Z}}(L) \cdot J \neq \mathrm{Sym}_{\mathbf{Z}}(L) \cdot J'$, there are neighborhood V_J and $V_{J'}$ of J and J' such that $\mathrm{Sym}_{\mathbf{Z}}(L) \cdot V_J \cap \mathrm{Sym}_{\mathbf{Z}}(L) \cdot V_{J'} \neq \phi$. The main point here is that for two relatively compact subsets X and Y of $\mathrm{Reas}(W)$ the set $Z = Z(X, Y) = \{A \in \mathrm{Sym}_{\mathbf{Z}}(L) \mid X \cap A \cdot Y \neq \phi\}$ is finite. This follows because by Theorem 7.1 $\mathrm{Sym}_{\mathbf{R}}(2g)/U(g) \approx \mathrm{Reas}(W)$. Let $\sigma : \mathrm{Sym}_{\mathbf{R}}(2g)$ be the projection $\mathrm{Sym}_{\mathbf{R}}(2g) \to \mathrm{Reas}(W)$. As the unitary group $U(g)$ is compact, it follows that $\sigma^{-1}(\overline{X})$ and $\sigma^{-1}(\overline{Y})$ are compact. Now $Z \subset \mathrm{Sym}_{\mathbf{Z}}(L) \cap \{\sigma^{-1}(\overline{X}) \cdot (\sigma^{-1}(\overline{Y}))^{-1}\}$ is the intersection of a compact subset with a discrete subset. Hense Z is finite and the main point is verified.

For a) $G_J = Z(J, J)$ is finite. Let K be any relatively compact neighborhood of J. Then $Z(K, K)$ is finite and equal $G_J \coprod F$. Let $L = K \cap \bigcap_{f \in F} U_f \cap f^{-1}V_f$ where U_f and V_f are disjoint neighborhoods of J and $f * J$. It follows that $Z(L, L) = G_J$ and we may take $U_J = \bigcap_{g \in G_J} g * L$. For b) let K and K' be relatively compact neighborhoods of J and J'. Then $Z = Z(K, K')$ is finite.

For each f in Z let P_f and P'_f be disjoint neighborhoods of J and fJ'. Then $V_J = K \cap \cap P_f$ and $V_{J'} = K' \cap f^{-1}P'_f$ solves the problem. $\qquad\square$

Therefore if we give $\mathrm{Sym}_{\mathbf{Z}}(L) \setminus \mathrm{Reas}(W)$ the quotient topology by b), it is a Hausdorff space. By a) it has a quotient analytic space structure which is locally isomorphic to the quotient of a manifold by a finite group. Thus the moduli $M(L, W)$ is naturally a separated analytic space.

A naive mistake is to try to construct a family of polarized abelian varieties over $M(L, W)$. The guess is consider $\pi'' : \mathrm{Sym}_{\mathbf{Z}}(L) \setminus B/L \to M(L, W)$. The problem is the fiber over $X(J)$ is $X(J) \setminus \setminus \mathrm{Aut}(X(J))$ which is not an abelian variety. The remedy is to consider a smaller subgroup $\Gamma \subset \mathrm{Sym}_{\mathbf{Z}}(L)$ such that Γ has no fixed points in $\mathrm{Reas}(W)$. Then $\pi'' : \Gamma \backslash B/L \to \Gamma \backslash \mathrm{Reas}(W)$ is a family of principal polarized abelian variety over a smooth base. In Proposition 9.7 we will see that we make take $\Gamma = \mathrm{Ker}(\mathrm{Sym}_{\mathbf{Z}}(L) \to \mathrm{Aut}_{\mathbf{Z}/n\mathbf{Z}}(L/nL))$ where n is an integer ≥ 3. In this case $\Gamma \backslash \mathrm{Reas}(W)$ is moduli space of polarized abelian varieties with a kind of level n structure and it is a Galois covering of $M(L, W)$ with Galois group $\mathrm{Sym}_{\mathbf{Z}}(L)/\Gamma$ as Γ is a normal subgroup.

Next we want to study the action of $\mathrm{Sym}_{\mathbf{Z}}(L)$ on $\mathrm{Reas}(W)$ using coordinates.

Let $a_1, \ldots, a_g, b_1, \ldots, b_g$ be a canonical basis of L with elementary divisors $e_1, \ldots, e_g$. Let e be the diagonal matrix with coefficients $e_1, \ldots, e_g$. Then the skew-form E is given by $\begin{bmatrix} 0 & e \\ -e & 0 \end{bmatrix}$. An element of $\mathrm{Sym}_{\mathbf{Z}}(L)$ is a $2g \times 2g$ integral matrix $\begin{bmatrix} a & b \\ c & d \end{bmatrix}$ such that ${}^t\begin{bmatrix} a & b \\ c & d \end{bmatrix}\begin{bmatrix} 0 & e \\ -e & 0 \end{bmatrix}\begin{bmatrix} a & b \\ c & d \end{bmatrix} = \begin{bmatrix} 0 & e \\ -e & 0 \end{bmatrix}$. Let Γ_e denote this group. Let $b'_i = \frac{1}{e_i}b_i$. Then $a_1, \ldots, a_g, b'_1, \ldots, b'_g$ is a symplectic basis of W and as such it defines a bijection $R : \mathrm{Reas}(W) \to S_g$. We know how $\mathrm{Sym}_{\mathbb{R}}(W)$ acts on $B \approx \mathbb{C}^g \times S_g$ in terms of a block decomposition of $\mathrm{Sym}_{\mathbb{R}}(W)$ with respect to the second basis. If $\begin{bmatrix} a & b \\ c & d \end{bmatrix}$ is in Γ_e then the corresponding element of $\mathrm{Sym}_{\mathbb{R}}(W)$ is $\begin{bmatrix} a & be \\ e^{-1}c & e^{-1}de \end{bmatrix}$. Thus the corresponding action of B is given by $\begin{bmatrix} a & b \\ c & d \end{bmatrix} * (z, \tau) = ((c\tau + de)^{-1}ez, (a\tau + be)(c\tau + de)^{-1}e)$. So $M(L, W)$ is isomorphic to $\Gamma_e \setminus S_g$ with the action given by the second coordinate.

§ 7.4 Families of Ample Sheaves on a Variable Abelian Variety

In the situation of Section 7.3 we want to consider families of invertible sheaves on the basic family $\pi' : B/L \to \mathrm{Reas}(W)$. First of all we consider the multiplier $A_l(z) = \alpha_l e^{+\pi H_J(z,l)+\frac{\pi}{2}H_J(l,l)}$ where $\alpha_{l+l'} = \alpha_l \cdot \alpha_{l'}(-1)^{E(l,l')}$ as usual. For fixed

J this defines an ample invertible sheaf $\mathcal{M}_J$ on $X(J)$. This family does not depend holomorphically on J but only real analytically (recall in coordinates H_J is given by $(\operatorname{Im}\tau)^{-1}$ where $\tau = R(J)$). What we can do is change the horizontal structure on the family $\mathcal{M}_J$ so that it is holomorphic. This can be done in many ways depending on the choice of maximal isotropic subgroup A of L. This idea goes back to the proof of Theorem 2.1 in section 2.2.

For J in $\operatorname{Reas}(W)$, let $S_{A,J}$ be the symmetric complex bilinear form on (W, J) such that $S_{A,J} = H_J$ on $A \otimes_{\mathbb{Z}} \mathbb{R} \times A \otimes_{\mathbb{Z}} \mathbb{R}$. Let $\widehat{\;}(A, J) : (W, J) \to \mathbb{C}$ be the complex linear mapping such that $v\widehat{\;}(A, J)(w) = E(w, v)$ for all w in $A \otimes_{\mathbb{Z}} \mathbb{R}$. Assume that α is identically one on A. Then $K_{l,J}(v) = \alpha(l)e^{+\pi i l\widehat{\;}(A,J)(l)+2\pi i l\widehat{\;}(A,J)(v)}$ is factor of automorphy. Let $\mathcal{N}_{A,J}$ be the invertible sheaf on $X(J)$ such that multiplication by $e^{\frac{\pi}{2}S_{A,J}(v,v)}$ gives an isomorphism $\varphi_{A,J} : \mathcal{M}_J \to \mathcal{N}_{A,J}$.

Lemma 7.8. *$\mathcal{N}_A$ is an invertible sheaf on B/L. In other words the multipliers $K_{l,J}(v)$ are holomorphic for (v, J) in B.*

Proof. Let $a_1, \ldots, a_g$ be a basis of A. Then we choose a canonical basis $a_1, \ldots, a_g$, $b_1, \ldots, b_g$ of L. Let $a_1, \ldots, a_g, b_1', \ldots, b_g'$ be the associated symplectic basis of W. Then we have the standard isomorphism $\varphi : B \approx \mathbb{C}^g \times S_g$ and $L \approx \mathbb{Z}^g \oplus \mathbb{Z}^g = \oplus \mathbb{Z}a_j \oplus \oplus bZ(-b_j)$.

Claim. In terms of this isomorphism $K_{l,J}$ is given by

$$e^{-\pi i{}^t(\bar{e}l_2)\tau\bar{e}l_2 - 2\pi i{}^t(el_2)z}$$

where $l = l_1 \oplus l_2$ and $\tau = R(J)$ and z corresponding to v.

This is clear because $K_{l,J}(v)$ is one if l is in A by the calculation in the proof of Theorem 2.1. Thus $K_{l_1+l_2,J}(v) = K_{l_2,J}(v)$ which can be computed by Lemma 5.2 which determines $\widehat{\;}$ in terms of τ. $\qquad\square$

Next let $\widetilde{A}$ be another maximal isotropic subgroup of L such that $\alpha|_{\widetilde{A}} \equiv 1$. Then we have the isomorphism $K_{\widetilde{A}}^A \equiv \varphi_{\widetilde{A},J} \circ \varphi_{A,J}^{-1} : \mathcal{N}_{A,J} \to \mathcal{N}_{\widetilde{A},J}$.

Lemma 7.9. *$K_{\widetilde{A}}^A$ is an isomorphism $\mathcal{N}_A \xrightarrow{\approx} \mathcal{N}_{\widetilde{A}}$ of invertible sheaves on B/L.*

Proof. $K_{\widetilde{A}}^A$ is given by multiplication by $e^{-\frac{\pi}{2}(S_{\widetilde{A},J}(v,v)-S_{A,J}(v,v))}$. We need to see that the function $S_{\widetilde{A},J}(v,v) - S_{A,J}(v,v) \equiv Q_J(v)$ is $\mathbb{C}$-analytic in v and J simultaneously. We will do this in coordinates (z, τ) with respect to $a_1, \ldots, a_g$, $b_1, \ldots, b_g$. Let $\tilde{a}_1, \ldots, \tilde{a}_g, \tilde{b}_1, \ldots, \tilde{b}_g$ be another canonical basis of L such that

$\widetilde{A} = \oplus \mathbb{Z} a_j$. Thus we have a symplectic transformation $\begin{bmatrix} \alpha & \beta \\ \gamma & \sigma \end{bmatrix}$ in $\mathrm{Sym}_{\mathbb{Z}}(L)$ which takes the ordinary basis to the one with the $\sim$. $\square$

This result will prove Lemma 7.9.

Sublemma 7.10. $Q_J(v)$ *is given by* $-2i \, {}^t z \, {}^t \gamma \, {}^t(\gamma \tau + \delta e)^{-1} z$ *in terms of the complex basis* $a_1, \ldots, a_g$ *of* (W, J).

Proof. In this basis $S_{A,J}(v,v)$ is given by ${}^t z \, \mathrm{Im}(\tau)^{-1} z$ by the remarks in the proof of Theorem 5.8. Let $\tilde{\tau}$ be point in S_g corresponding to J computed with respect to $\tilde{a}_1, \ldots, \tilde{a}_g, \tilde{b}_1, \ldots, \tilde{b}_b$. Similarly $S_{\widetilde{A}, J}(v, v)$ is given by ${}^t \tilde{z} \, \mathrm{Im}(\tilde{\tau})^{-1} \tilde{z}$ with respect to the complex basis $\tilde{a}_1, \ldots, \tilde{a}_g$. We know that $\tilde{z} = e \, {}^t(\sigma \tau + \delta e)^{-1} z$ and $\tilde{\tau} = (\alpha \tau + \beta e)(\gamma \tau + \delta e)^{-1} e$. We also have the equation

$$(*) \qquad {}^t \tilde{z} \, \mathrm{Im}(\tilde{\tau})^{-1} \overline{\tilde{z}} = {}^t z \, \mathrm{Im}(\tau)^{-1} \bar{z}$$

as H_J is independent of the coordinates.

Now $Q_J(v) = {}^t \tilde{z} \, \mathrm{Im}(\tilde{\tau})^{-1} \tilde{z} - {}^t z \, \mathrm{Im}(\tau)^{-1} z$ which is given in the ordinary bases by the matrix ${}^t \{ e \, {}^t(\gamma \tau + \delta e)^{-1} \} \, \mathrm{Im}((\alpha \tau + \beta e)(\gamma \tau + \delta e)^{-1} e)^{-1} \{ e \, {}^t(\gamma \tau + \delta e)^{-1} \} - (\mathrm{Im}\, \tau)^{-1}$. By $(*)$ we have

$${}^t \{ e \, {}^t(\gamma \tau + \delta e)^{-1} \} \, \mathrm{Im}((\alpha \tau + \beta e)(\gamma \tau + \delta e)^{-1} e)^{-1} \{ e \, {}^t(\gamma \bar{\tau} + \delta e)^{-1} e \} = (\mathrm{Im}\, \tau)^{-1} \, .$$

So $Q_J(v)$ is given by

$$\begin{aligned}
{}^t \{ e \, {}^t(\gamma \tau + \delta e)^{-1} \} \, &{}^t \{ e \, {}^t(\gamma \tau + \delta e)^{-1} \}^{-1} (\mathrm{Im}\, \tilde{\tau})^{-1} (e \, {}^t(\gamma \bar{\tau} + \delta)^{-1})^{-1} \{ e \, {}^t \\
&\times (\gamma \tau + \delta e)^{-1} e \} - (\mathrm{Im}\, \tau)^{-1} \\
&= (\mathrm{Im}\, \tau)^{-1} [\, {}^t(\gamma \bar{\tau} + \delta) \, {}^t(\gamma \tau + \delta e)^{-1} - 1] \\
&= (\mathrm{Im}\, \tau)^{-1} [\, {}^t(\gamma \bar{\tau} + \delta) - {}^t(\gamma \tau + \delta e)] \, {}^t(\gamma \tau + \delta e)^{-1} \\
&= (\mathrm{Im}\, \tau)^{-1} [-2i \, {}^t(\gamma \, \mathrm{Im}\, \tau) \, {}^t(\gamma \tau + \delta e)^{-1}] = -2i \, {}^t \gamma \, {}^t(\gamma \tau + \delta e)^{-1} \, . \qquad \square
\end{aligned}$$

Now $\mathscr{N}_A$ has a Hermitian metric so that $\varphi_{A,J}$ is an isometry for all J. Therefore $K_{\widetilde{A}}^A$ is an isomorphism of Hermitian invertible sheaves on B/L. We also note that we trivially have

$$K_{\widetilde{A}}^{\widetilde{\widetilde{A}}} \cdot K_{\widetilde{A}}^A = K_{\widetilde{\widetilde{A}}}^A$$

when these isomorphisms are defined.

Exercise 1. Give a formula for the metric on $\mathscr{N}_A$.

§ 7.5 Group Actions on the Families of Sheaves

We continue with the situation of the last section. For all J in $\mathrm{Reas}(W)$, $K(\mathscr{M}_J) = K(\mathscr{N}_{A,J})$ is independent of J. In fact it is $K = L^\wedge/L$. Let $A(E) = A \otimes \mathbb{R} \cap L^\wedge/L$ depend on the choice of the maximal isotropic subgroup A of L.

For the next result we need a definition. Let X be a complex manifold on which a group Γ acts by biholomorphic mappings. Let $\mathscr{L}$ be a coherent sheaf on X. An action of Γ on $\mathscr{L}$ is the following data; for each element γ of Γ we have an isomorphism $\alpha_\gamma : \gamma * \mathscr{L} \xrightarrow{\approx} \mathscr{L}$ such that for any pair γ_1 and γ_2 of elements we have a commutative diagram

$$
\begin{array}{ccc}
\alpha_{\gamma_1 \cdot \gamma_2} : (\gamma_1 \cdot \gamma_2)^* \mathscr{L} & \longrightarrow & \mathscr{L} \\
\| & & \uparrow \alpha_{\gamma_2} \\
(\gamma_2)^*(\gamma_1 * \mathscr{L}) & \xrightarrow{\gamma_2^*(\alpha_{\gamma_1})} & \gamma_2^* \mathscr{L}
\end{array}
$$

or simple $\alpha_{\gamma_1 \cdot \gamma_2} = \alpha_{\gamma_2} \circ \gamma_2^*(\alpha_{\gamma_1})$.

If $\mathscr{L} = \mathcal{O}_X$, α_γ is multiplication by a no-where zero holomorphic function $A_\gamma(x)$ on X such that the factors of automorphy equation $A_{\gamma_1 \cdot \gamma_2}(x) = A_{\gamma_2}(x) A_{\gamma_1}(\gamma_2 \cdot x)$ holds.

Now for each J we have an action of $A(E)$ on $\mathscr{N}_{A,J}$ given by transcribing via the automorphism $\varphi_{A,J} : \mathscr{M}_J \to \mathscr{N}_{A,J}$ the action of $A(E)$ on $\mathscr{M}_J$ via $f(v) \mapsto e^{-\pi H_J(\tilde{a}, \tilde{a}) \frac{\pi}{2} H_J(\tilde{a}, \tilde{a})} f(v + \tilde{a})$ where $\tilde{a}$ in $L^\wedge$ represents a. We will next check that this action depends holomorphically on J.

Lemma 7.11. *We have a natural action of $A(E)$ on $\mathscr{N}_A$ via isometries.*

Proof. Let $a_1, \ldots, a_g, b_1, \ldots, b_g$ be a canonical basis of L such that $A = \oplus \mathbb{Z} a_j$. Thus an element of $A(E)$ has the form $a = \sum \frac{n_j}{e_j} a_j$ where the n_j are integers. We need to compute the operation on $\mathscr{N}_{A,J}$ given by a where $R(J) = \tau$. Let g be a section of $\mathscr{N}_{A,J}$. By definition a takes this to the section

$$
g'(u) = e^{-\frac{\pi}{2} S_j(u,u)} e^{+\frac{\pi}{2} S_J(u+a, u+a)} g(u + a) e^{-\pi H_J(a,a) - \frac{\pi}{2} H_J(\tilde{a}, \tilde{a})} = g(u + a)
$$

because $S_J(-, a) = H_J(-, a)$. Thus this action is independent of J. $\qquad\square$

We apparently did something trivial. To do something less trivial, let $L = A \oplus B$ be a decomposition of L with respect to E. Assume that $\alpha|_B \equiv 1$. We have a similar group $B(E)$ of K which acts on $\mathscr{N}_{B,J}$ for all J. Similarly we have

Lemma 7.12. *We have a natural action of $B(E)$ on $\mathcal{N}_A$ via isometries.*

Proof. Now we assume that $B = \oplus \mathbb{Z} b_j$. A typical element $B(E)$ is represented by $b = \sum \frac{n_j}{e_j} b_j$ where the n_j are integers. As before we need to see that $-S_J(u, u) + S_J(u + b, u + b) - 2H_J(u, b) - H_J(b, b)$ is holomorphic in J and u. The above expression equals

$$+2S_J(u, b) + S_J(b, b) - 2H_J(u, b) - H_J(b, b) = -2i(2\widehat{b}_{J,A}(u) + \widehat{b}_{J,A}(b))$$
$$= -2i(2{}^t nu + {}^t n\tau n) \, .$$

So the action of b sends $g(u) \mapsto e^{-2\pi i {}^t nu - \pi i {}^t n\tau n} g(u - \tau e^{-1} n)$. $\qquad\square$

Now let $H(A, B) = \mathbb{C}^* \times A(E) \times B(E)$ with the usual multiplication. We have a natural action of $H(A, B)$ on $\mathcal{N}_A$. Thus we have an action of $H(A, B)$ on the direct image $\pi'_* \mathcal{N}_A$ which is a sheaf on $\mathrm{Reas}(W)$ where $H(A, B)$ acts trivially.

In other words we have an action of $H(A, B)$ on each fiber $\pi'_* \mathcal{N}_{A,J} = \Gamma(X(J), \mathcal{N}_{A,J})$ which varies analytically with J. In fact for each J, $H(A, B)$ acts by the theta group $H(\mathcal{N}_{A,J})$.

There is another group acting on $\mathcal{N}_A$. Let Γ_α be the subgroup of $\mathrm{Sym}_{\mathbb{Z}}(L)$ consisting of elements which preserve α on L.

Lemma 7.13. *We have a natural action of Γ_α on $\mathcal{N}_A$ by isometries.*

Proof. Let σ be in Γ_α. Then $\gamma^* \mathcal{N}_A = \mathcal{N}_{\sigma^{-1}(A)}$. The action is given by $K_A^{\sigma^{-1}A} = \mathcal{N}_{\sigma^{-1}(A)}$. To check that this is group action we need to see that

$$K_A^{(\sigma_1, \sigma_2)^{-1}A} = K_A^{\sigma^{-1}A} \circ (\sigma_2)^* K_A^{\sigma_1^{-1}A}$$

but

$$\sigma_2^* K_A^{\sigma_1^{-1}A} = K_{\sigma_2^{-1}A}^{\sigma_2^{-1}(\sigma_1^{-1}A)} = K_{\sigma_2^{-1}A}^{(\sigma_1\sigma_2)^{-1}A} \, .$$

So the result follows because of the remark at the end of the last section. Everything is an isometry because $\sigma^* \mathcal{M}_J = M_{\sigma^{-1}J}$. $\qquad\square$

Taking direct image we have

Corollary 7.14. *We have a natural action of Γ_α on the sheaves $\pi'_* \mathcal{N}_A \equiv Q_A$ on $\mathrm{Reas}(W)$ by isometries.*

Proof. The sheaf $\pi'_* \mathcal{N}_A$ is a locally free sheaf on $\mathrm{Reas}(W)$ by Theorem 3.14. Its metric is given by integration of the inner product in $\mathcal{N}_A$ over the fibers of π' with respect to normalized relative Haar measure. The corollary follows directly from the properties of integration. $\qquad\square$

Now we have action of the two groups Γ_α and $H(A, B)$ on $\mathcal{N}_A$. We may ask "How are they related"?

Lemma 7.15. *Γ_α normalizes the action of $H(A, B)$. Hence we have a semi-direct product $H(A, B) \times \Gamma_\alpha$ acting on $\mathcal{N}_A$.*

Proof. Let $m = (\lambda, a, b)$ be an element of the compact group $H_C(A, B)$ where $|\lambda| = 1$. Let p be an element of Γ_α. We want to show that $(pmp^{-1})^* = m'^*$ for some element m' of $H_C(A, B)$. Now $(pmp^{-1})^* = p^{*-1}m^*p^*$ is the composition of an isomorphism $\mathcal{N}_{A,J} \to \mathcal{N}_{A,p^{-1}J}$ with a covering of translation by $a + b$ and the inverse $\mathcal{N}_{A,p^{-1}J} \to \mathcal{N}_{A,J}$. Thus $(pmp^{-1})^*_J$ is a covering of translation by $p(a + b)$ and, hence has the form $(\mu(J), c, d)^*$. As the above isomorphisms are all isometries, we must have $|\mu(J)| = 1$. Now μ is an analytic function of J. Therefore it is constant. $\qquad\square$

Chapter 8. Modular Forms

§ 8.1 The Definition

We can write a canonical invertible sheaf on space $\mathrm{Reas}(W)$ as a fixed real
vector space W of dimension $2g$. We recall that we have the homomorphic
bundle $B \to \mathrm{Reas}(W)$ on rank g. One simply considers the higher exterior
power $\Lambda^g B$. The fiber of $\Lambda^g B$ at J is canonically isomorphic to the g-exterior
power of the tangent space of the abelian variety $X(J)$ at the origin. As B is
homogeneous with respect to the action of $\mathrm{Sym}_{\mathbb{R}}(V)$ so is $\Lambda^g B$. Recalling that
B is Hermitian we see that $\Lambda^g B$ is Hermitian in an invariant way but this is
not the interesting metric on $\Lambda^g B$.

We will define the related sheaves. Let $\mathcal{H}$ be the sheaf on $\mathrm{Reas}(J)$ such
that a section $\mathcal{H}$ over U is a holomorphic g-form ω on $\pi^{-1}U$ which is invariant
on the fibers of π. Let γ be an element of $\mathrm{Sym}_{\mathbb{R}}(W)$. Then γ^* of differential
forms defines an isomorphism of $\gamma^* \mathcal{H} \to \mathcal{H}$ which gives a (continuous) action
of the group $\mathrm{Sym}_{\mathbb{R}}(W)$ on $\mathcal{H}$ by the chain rule.

The interesting metric on $\mathcal{H}$ assumes a given lattice L in W where E is
integral on $L \times L$. We identify a section of $\mathcal{H}$ with a differential form on B/L in
the obvious way. The interesting metric is given by $(\omega_1, \overline{\omega}_2)_J = \frac{1}{(2i)^g} \int_{X(J)} \omega_1 \wedge
\overline{\omega}_2$. By the change of variables in integrals, the group $\mathrm{Sym}_{\mathbb{Z}}(L)$ acts on $\mathcal{H}$ by
isometries in this metric.

Another definition is as follows. A modular form for a group Γ in $\mathrm{Sym}_{\mathbb{R}}(L)$
of weight k is a global section ω of $\mathcal{H}^{\otimes k}$ which is invariant under Γ; i.e. $\gamma^* \omega = \omega$
for all γ in Γ. If $g = 1$ one normally requires a growth condition at the cusps
which I will ignore. Clearly the ring of modular forms of all weights forms a
graded ring. Unfortunately these rings have not yet been determined explicitly
(this is the first time we encountered abstract mathematics in this book).

In this section I want to write the notions in terms of coordinates for
comparison with the classical notation.

Let $a_1, \dots, a_g, b'_1, \dots, b'_g$ be a symplectic basis for W. We have the bijection
$R : \mathrm{Reas}(W) \to S_g$. Then $\pi : B \to \mathrm{Reas}(W)$ is just the projection on the second
factor $\pi : \mathbb{C}^g \times S_g \to S_g$. Thus a section of H over an open subset $\pi^{-1}(U)$ is
an expression $\omega = f(\tau)dz_1 \wedge \dots \wedge dz_g$ where f is a holomorphic function of τ on

U. Thus action of $\begin{bmatrix} \alpha & \beta \\ \alpha & \delta \end{bmatrix}$ in $\mathrm{Sym}_{\mathbb{R}}(W)$ sends ω into $\gamma^*\omega = f(\alpha\tau + \beta/(\gamma\tau + \delta))\det(\gamma\tau + \delta)^{-1}dz_1\wedge\ldots\wedge dz_g$ because $\gamma(z,\tau) = ((\gamma\tau + \delta)^{-1}z, (\alpha\tau + \beta)(\gamma\tau + \delta)^{-1})$. Thus a modular form of weight k is an expression

$$f(\tau)(dz_1\wedge\ldots\wedge dz_g)^{\otimes k}$$

where f is a holomorphic function of S_g satisfying $f((\alpha\tau + \beta)(\gamma\tau + \delta)^{-1}) = f(\tau)\det(\gamma\tau + \delta)^k$ for all $\begin{bmatrix} \alpha & \beta \\ \gamma & \delta \end{bmatrix}$ in Γ. Such a function is called automorphic.

Let $a_1,\ldots,a_g$, $b_1,\ldots,b_g$ be a canonical basis of L. Let $b'_i = \frac{1}{e_i}b_i$ as usual. Let $\omega_1 = f(\tau)dz_1\wedge\ldots\wedge dz_g$ and $\omega_2 = g(\tau)dz_1\wedge\ldots\wedge dz_g$ be two sections of $\mathscr{H}$. Then we want to compute their inner product.

Lemma 8.1. $(\omega_1,\omega_2)_\tau = \frac{1}{2^g}f(\tau)\overline{g(\tau)}(h(X(J), H_J, A))^2$ *where* $R(J) = \tau$ *and* $A = \bigoplus_j \mathbb{Z}a_j$.

Proof. By definition $h(X(J), H(J), A) = \sqrt{2^g(\pm dz_1\wedge\ldots\wedge dz_g, \pm dz_1\wedge\ldots\wedge dz_g)}$. $\square$

Thus the geometric height gives the metric in $\mathscr{H}$.

§ 8.2 The Realtionship Between $\pi'_*\mathscr{N}_A$ and H in the Principally Polarized Case

In this section we assume that $|\det E| = 1$. Thus $\pi'_*\mathscr{N}_A$ is an invertible sheaf. In the situation of Section 7.5 we have a group action of Γ_α by isometries on $Q(A) \equiv \pi'_*\mathscr{N}_A$ and by the last section we have an action $\Gamma_\alpha \subset \mathrm{Sym}_{\mathbb{Z}}(L)$ on $\mathscr{H}$ with respect to metric defined by L. The obvious question is "Is there any relation between these two actions?".

The key to understanding this problem is

Lemma 8.2. *Let $\mathscr{L}$ be an invertible sheaf with a Hermitian structure on a connected complex manifold X. Let Γ be a group acting holomorphically on X. Then any two actions of Γ on $\mathscr{L}$ which acts by isometries differ by a character* $\chi : \Gamma \to U(1)$.

Proof. Let $*$ and $**$ be the two actions. Then for each γ in Γ we have two isometries γ^*, $\gamma^{**} : \gamma^*\mathscr{L} \xrightarrow{\approx} \mathscr{L}$. By Lemma 4.7 we have a unitary constant $\chi(\gamma)$ such that $\chi(\gamma)\gamma^* = \gamma^{**}$. As $*$ and $**$ are group actions, χ is a character of Γ. $\square$

To use this result in practice, let $\mathcal{L}_1$ and $\mathcal{L}_2$ be two invertible sheaves with Hermitian structures on X. Assume that both $\mathcal{L}_1$ and $\mathcal{L}_2$ are given Γ-actions. Then if $\alpha : \mathcal{L}_1 \xrightarrow{\approx} \mathcal{L}_2$ is an isometry, there is a unitary character χ of $\mathcal{L}$ such that α is isomorphism between sheaves with Γ-action where the action of Γ on $\mathcal{L}_1$ is twisted by χ.

We will apply this principle when $X = \mathrm{Reas}(W)$, $\mathcal{L}_1 = Q(A)^{\otimes 2}$ and $\mathcal{L}_2 = \mathcal{H}^{\otimes -1}$. To define the isometry $\alpha : Q(A)^{\otimes 2} \to \mathcal{H}^{\otimes -1}$ we use a canonical basis $a_1, \ldots, a_b, b_1, \ldots, b_g$ of L. Define α in coordinates by $\alpha(f(\tau)\eta) = f(\tau)\frac{\partial}{\partial z_1} \wedge \ldots \wedge \frac{\partial}{\partial z_g} \frac{2^{g/2}}{\det(E)^{1/8}}$. By Theorem 5.9, this is an isometry. Thus we get

Theorem 8.3. *There is a character* $\chi : \Gamma_\alpha \to U(1)$ *such that we have an isometry of sheaves with* Γ_γ*-action*

$$\alpha : Q(A)^{\otimes 2}[\chi] \longrightarrow \mathcal{H}^{\otimes -1}$$

where $[\,]$ *denotes twisting the* Γ_γ*-action by the character* χ.

With this theorem in hand all of our previous questions boil down to the single one, "What is χ?". We can give an elementary result.

Lemma 8.4. *Let* γ *be an element of* Γ_α *such that* $\gamma(A) = A$. *Then* $\chi(v) = \det(\gamma|_A)$.

Proof. In coordinates γ is given by $\begin{bmatrix} \alpha & \beta \\ 0 & \delta \end{bmatrix}$ where ${}^t\alpha e \delta = e$. So $\det(\gamma|_A) = \det(\alpha) = \det(\delta^{-1})$. Now $\gamma^* dz_1 \wedge \ldots \wedge dz_g = \det(\delta^{-1}) dz_1 \wedge \ldots \wedge dz_g$ because $\gamma \times z = e\, {}^t(\delta e)^{-1} z$. Thus $\gamma^* \frac{\partial}{\partial z_1} \wedge \ldots \wedge \frac{\partial}{\partial z_g} = \det(\gamma|_A)\frac{\partial}{\partial z_1} \wedge \ldots \wedge \frac{\partial}{\partial z_g}$. To prove this lemma it is enough to show that $\gamma^*\eta = \eta$ or, what is the same as $\gamma^*\eta(\delta_0) = \eta(\delta_0)$. One easily checks that $\eta(\delta_0)$ depends only on A not B. $\square$

Let $\Gamma_{\alpha,A}$ be the subgroup of all such α as in the lemma. Let $\Gamma(n)$ be the kernel of $\mathrm{Sym}_{\mathbb{R}}(L) \to \mathrm{GL}_{\mathbb{Z}/n\mathbb{Z}}(L/nL)$. Then trivially we have $\Gamma(4) \subset \Gamma(2) \subset \Gamma_\alpha$. In the next section we will prove

Lemma 8.5.
   a) Γ_α *is generated by* $\Gamma_{\alpha,A}$ *together with square roots of elements of* $\Gamma_{\alpha,A}$.
   b) $\Gamma(2)$ *is contained in the group spanned by conjugates of* $\Gamma_{\alpha,A}$ *in* Γ_α.
   c) $\Gamma(4)$ *is contained in the group spanned by conjugates of* $\mathrm{Ker}\,\chi|_{\Gamma_{\alpha,A}}$ *in* Γ_α.

Right now we can conclude

Theorem 8.6.
  a) $\chi^4 \equiv 1$,
  b) $\chi^2|_{\Gamma(2)} \equiv 1$, *and*
  c) $\chi|_{\Gamma(4)} \equiv 1$.

Proof. This follows immediately from the two Lemmas as χ is a character. $\square$

One should notice that $\Gamma_\alpha = \mathrm{Sym}_{\mathbf{Z}}(L)$ if $\tilde{e}$ is even.

Corollary 8.7. *We have canonical isomorphism*
  a) $Q(A)^{\otimes 8} \to \mathcal{H}^{\otimes -4}$ *as Γ_α-metric sheaves,*
  b) $Q(A)^{\otimes 4} \to \mathcal{H}^{\otimes -2}$ *as $\Gamma(2)$-metric sheaves,*
  c) *and upto sign* $Q(A)^{\otimes 2} \to \mathcal{H}^{\otimes -1}$ *as $\Gamma(4)$-metric sheaves.*

Proof. The isometry α of Theorem 8.3 is determined upto sign by A. The canonical statement follows and for the rest see Theorem 8.3 and 8.6. $\square$

Next we will write the classical functional equation of the theta function in our language. There is a homomorphism of sheaves

$$\beta : Q(A) \longrightarrow \pi' \mathcal{N}_A \text{ given by } \beta(\lambda) = \lambda(\delta_0) \ .$$

From the definitions β is equivariant for the action Γ_γ on the two sheaves. Thus we have an equivariant mapping $\gamma = \beta^2 \circ \alpha^{-1}[\chi^{-1}] : \mathcal{H}^{-1} \to \pi'_* \mathcal{N}_A[\chi^{-1}]$. Let $|_0$ be evolutions at 0, $\pi'_* \mathcal{N}_A \to \mathcal{O}_{\mathrm{Reas}(W)}$ which is equivariant for the natural Γ_α action by pull-back on $\mathcal{O}_{\mathrm{Reas}(W)}$. Thus $(|_0 \circ \gamma)^{\mathrm{dual}} : \mathcal{O}_{\mathrm{Reas}(W)} \to \mathcal{H}(\chi)$ is Γ_γ-equivariant. Let M be the image of the function 1 is Γ_α-invariant. Then M is a modular form twisted by χ for the group Γ_α but M is essentially $(\delta_0|_0)^2 dz_1 \wedge \ldots \wedge dz_g$.

§ 8.3 Generators of the Relevant Discrete Groups

Let Γ be the standard integral symplectic group of the usual form E consisting of integral $2g \times 2g$ matrices of the form $\begin{bmatrix} \alpha & \beta \\ \gamma & \delta \end{bmatrix}$ where

$$^t\begin{bmatrix} \alpha & \beta \\ \gamma & \delta \end{bmatrix} \begin{bmatrix} 0 & 1 \\ -1 & 0 \end{bmatrix} \begin{bmatrix} \alpha & \beta \\ \gamma & \delta \end{bmatrix} = \begin{bmatrix} 0 & 1 \\ -1 & 0 \end{bmatrix} \ .$$

Let $L = \mathbf{Z}^{2g}$ be the lattice on which Γ operates naturally. A vector in L is given by $(x_1, \ldots, x_g, y_1, \ldots, y_g)$ in coordinates.

Let n be a positive integer. Let $\Gamma(n)$ be the subgroup of Γ of matrices of the form $\begin{bmatrix} 1+n\alpha & n\beta \\ n\gamma & 1+n\delta \end{bmatrix}$ where the α, β, γ and δ are integral. Let Γ' be the group of similar elements of the form $\begin{bmatrix} 1+4\alpha & 2\beta \\ 2\gamma & 1+4\delta \end{bmatrix}$. We have inclusions

$$\Gamma(4) \subset \Gamma' \subset \Gamma(2) \subset \Gamma .$$

We need another group $\Gamma(1,2)$ which consists of all elements $\begin{bmatrix} \alpha & \beta \\ \gamma & \delta \end{bmatrix}$ of Γ which preserves the quadratic form $Q(x,y) = \sum_{j=1}^{g} x_j y_j$ modulo 2. Clearly $\Gamma(2) \subset \Gamma(1,2) \subset \Gamma$. We want to find generators of the groups Γ', $\Gamma(2)$, $\Gamma(1,2)$ and Γ.

For $1 \leq j \leq g$ let T_g be the metric of the transformation of L which sends x_j to y_j and y_j to $-x_j$ and leaves the other coordinates fixed. What we intend to prove is

Theorem 8.8.

a) Γ' is generated by its elements of the form $\begin{bmatrix} 1 & 0 \\ 2B & 1 \end{bmatrix}$ and $\begin{bmatrix} 1 & 2B \\ 0 & 1 \end{bmatrix}$ where B is integral and symmetric.

b) $\Gamma(2)$ is generated by its elements of the form $\begin{bmatrix} 1 & 2B \\ 0 & 1 \end{bmatrix}$, $\begin{bmatrix} 1 & 0 \\ 2B & 1 \end{bmatrix}$ and $\begin{bmatrix} A & 0 \\ 0 & {}^tA^{-1} \end{bmatrix}$ where B is integral and symmetric and $A = 1 + 2C$ where C is integral.

c) $\Gamma(1,2)$ is generated by $T_1,\ldots,T_g$ and its elements of the form $\begin{bmatrix} A & 0 \\ 0 & {}^tA^{-1} \end{bmatrix}$ and $\begin{bmatrix} 1 & B \\ 0 & 1 \end{bmatrix}$ where B is symmetric and has even diagonal.

d) Γ is generated by $T_1,\ldots,T_g$ and its elements of the form $\begin{bmatrix} A & 0 \\ 0 & {}^tA^{-1} \end{bmatrix}$ and $\begin{bmatrix} 1 & B \\ 0 & 1 \end{bmatrix}$ where B is symmetric.

Remark. d) may be proven easier directly but we will prove these statements in order.

Proof. We begin with a modification of the usual procedure for finding the greatest common divisor of two integers.

Lemma 8.9. *Let x be an odd number and y be an even integer. After a finite number of steps using transformations $S_1 : x \to x$ and $y \pm 2x$ or $S_2 : y \to y$ and $x \to x \pm 2y$, we may reduce (x,y) to $(\pm d, 0)$ where $d = \gcd(x,y)$.*

Proof. Let (x', y') be equivalent to (x, y) under the group generated by S_1 and S_2 such that $|x'| + |y'|$ is minimal. Clearly as y' is even and x' is odd, $|x'| < |y'|$ if $y' \neq 0$ and $|y'| < |x'|$. Thus $y' = 0$. Also we note that $\gcd(x, y)$ is invariant under the group. So the result follows. $\square$

To prove a) by an easy induction the result will follow if we can prove the following; given two vectors $(x_1, , \ldots, x_g, y_1, \ldots, y_g) = K_1$ and $(u_1, \ldots, u_g, v_1, \ldots, v_g) = K_2$ in L such that $x_1 \equiv 1 \equiv v_1$, $x_2 \equiv v_2 \equiv x_g \equiv 0((4))$ the other entrees are even, and $\sum x_j v_j - \sum y_j u_j = 1$ there is a transformation in the groups generated by $\begin{bmatrix} 1 & 0 \\ 2B & 1 \end{bmatrix}$ and $\begin{bmatrix} 1 & 2B \\ 0 & 1 \end{bmatrix}$ where B is integral and symmetric such that K_1 and K_2 are transformed to the unit vectors e_1 and e_{g+1}.

First if $2 \leq i \leq g$ we use the transformation

$$x_1, x_i, y_1, y_i \longrightarrow x_1 \pm 2y_i, x_i \pm 2y_1, y_1, y_i \text{ or } x_1, x_i, y_1 \pm 2x_i, y_i \pm 2x_1$$

until by the lemma $y_2 = \ldots = y_g = 0$. Then apply $W : x_1, y_1 \mapsto x_1 \pm 2y_1, y_1$ or $x_1, y_1 \pm 2x_1$ until $y_1 = 0$. Next we want to make a series of transformations so that x_1 divides $x_2, \ldots, x_g, y_1, \ldots, y_g$. During this process we will preserve the condition $x_1 | y_1$ for $2 \leq j \leq g$. For $2 \leq j \leq g$ we apply $x_1, x_j, y_1, y_i \to x_1, x_j, y_1 + 2x_i, y_i + 2x_1$ so that $y_1 = 2x_i$. Then we apply process W to make $y_1 = 0$. Then by the lemma $x_1 | 2x_i$ and thus $x_1 | x_i$ as x_1 is odd. Therefore we have x_1 divides all of the x_j and y_j. On the other hand as $(K_1, K_2) = 1$, $\gcd(\{x_j, y_j\}) = \pm 1$. Thus $x_1 = \pm 1$ or, hence, $x_1 = 1$ as $x_1 \equiv 1(4)$.

Next we repeat the procedure in the beginning so that $y_1, \ldots, y_g$ are zero. Then by a W-process we make $y_1 = 2$. Then we apply $x_1, x_i, y_1, y_i \to x_1 \pm 2y_i, x_i \pm 2y_1, y_1, y_i$ to make y_i zero as $y_i \equiv 0(4)$. Finally we apply W to make y_1 zero again. Thus K_1 is now e_1. So $v_1 = 1$ as $E(K_1, K_2) = 1$. We want to change K_2 to e_2 without changing e_1. If $g = 2$, make $u_3 = \ldots = u_g = 0$ and $u_2 = 2$ by $U : x_1, x_i, y_1, y_i \to x_1 \pm 2y_i, x_i \pm 2y_1, y_1, y_i$. Then make $v_2 = \ldots = v_g = 0$ by $x_2, x_i, y_2, y_i \to x_2, x_i, y_2 \pm 2x_i, y_i \pm 2x_2$. Then make u_2 equal zero by U transformation. Then if $g \geq 1$ just make $u_1 = 0$ by Ws. This accomplishes the task. This proves a).

To prove b) we have an injection $i : \Gamma(2)/\Gamma(4) \to \mathrm{Sym}_{\mathbf{Z}/4\mathbf{Z}}(L \otimes \mathbf{Z}/4\mathbf{Z})$. The image of i is contained in the matrices of the form $\begin{bmatrix} 1 + 2\alpha & 2\beta \\ 2\gamma & 1 + 2\delta \end{bmatrix}$ where $\alpha, \beta, \gamma, \delta$ are $g \times g$ matrices with coefficients in $\mathbf{Z}/2\mathbf{Z}$ which preserves $E \otimes \mathbf{Z}/4\mathbf{Z}$. Thus $-\,{}^t\delta = \alpha$ and γ and β are symmetric and the group law for these matrices is addition of the α, β, γ and δ. We will show that the proposed generators of $\Gamma(2)$ generates these subgroups. The $\begin{bmatrix} 1 & 2\beta \\ 0 & 1 \end{bmatrix}$ and $\begin{bmatrix} 1 & 0 \\ 2\gamma & 1 \end{bmatrix}$ are obvious in the image. For $\begin{bmatrix} 1 + 2\alpha & 0 \\ 0 & 1 - {}^t2\alpha \end{bmatrix}$ by addition we may assume that α is upper or lower diagonal or diagonal with only one in the ith-place. The first two cases are given by $\begin{bmatrix} A & 0 \\ 0 & {}^tA^{-1} \end{bmatrix}$ in an obvious way. For the other just take A to be the diagonal matrix with ones everywhere except for a minus one in the ith-place.

The proof of c) is similar. Consider the injection $i : \Gamma(1,2)/\Gamma(2) \to O_{\mathbb{Z}/2\mathbb{Z}}(L \otimes \mathbb{Z}/2\mathbb{Z})$ where O is the orthogonal group of $Q \otimes \mathbb{Z}/2\mathbb{Z}$. One shows that the image of the would-be generators generate the orthogonal group. While we are at it we can do the same thing with d). Here we need to show that the generators generate $\mathrm{Sp}_{\mathbb{Z}/2\mathbb{Z}}(L \otimes \mathbb{Z}/2\mathbb{Z})$. Thus we need to find generators over a field $F = \mathbb{Z}/2\mathbb{Z}$ in this case. What we need is

Lemma 8.10. *a)* $O_F(Q \otimes F)$ *is generated by the image of* $T_1, \ldots, T_g,$ $\begin{bmatrix} A & 0 \\ 0 & {}^tA^{-1} \end{bmatrix}$ *where A is either a permutation matrix or* $1 + F_{ij}$, $i \neq j$, *where F_{ij} is the matrix with only ones in the (i,j) place and* $\begin{bmatrix} 1 & F_{ij} + F_{ji} \\ 0 & 1 \end{bmatrix}$ *where $i \neq j$.*

b) $\mathrm{Sp}(L \otimes \mathbb{Z}/2\mathbb{Z})$ *is generated by the same plus* $\begin{bmatrix} 1 & F_{ij} \\ 0 & 1 \end{bmatrix}$ *for all i.*

Proof. As char $F = 2$, E is the bilinear form associated to Q. By the obvious induction on g to prove a) we need to show that if we have two vectors $K_1 = (x_1, \ldots, x_g, y_1, \ldots, y_g)$ and $K_2 = (u_1, \ldots, u_g, v_1, \ldots, v_g)$ in $L \otimes \mathbb{Z}/2\mathbb{Z}$ such that $E(K_1, K_2) = 1$ and $Q(K_1) = Q(K_2) = 0$ we can transform them to e_1 and e_{g+1} using our generators. For b) we drop the condition on Q but have more generators at our disposal.

Using permutations and T's we may assume that $x_1 = 1$. Using $x_1, x_j, y_1, y_j \to x_1, x_j + x_1, y_1 + y_j, y_j$ we may assume that $x_2 = \ldots = x_g = 0$. Thus as $Q(K_1) = 0$ we have $y_1 = 0$. Now $T_1 \ldots T_g = \begin{bmatrix} 0 & 1 \\ 1 & 0 \end{bmatrix}$ and $\begin{bmatrix} 0 & 1 \\ 1 & 0 \end{bmatrix} \begin{bmatrix} 0 & 1 \\ 1 & 0 \end{bmatrix}$ $\begin{bmatrix} 1 & F_{ij} + F_{ji} \\ 0 & 1 \end{bmatrix} = \begin{bmatrix} 0 & 1 \\ 1 & 0 \end{bmatrix} = \begin{bmatrix} 1 & 0 \\ F_{ij} + F_{ji} & 1 \end{bmatrix}$. Thus we may use the transforms $x_1, x_j, y_1, y_j \to x_1, x_j, y_1 + x_j, y_j + x_1$ $i = 1$, $j = 2, \ldots, g$, we can assume that $y_2 = \ldots = y_g = 0$. Thus K_1 is now e_1 and, hence, $v_1 = 1$. As $Q(K_2) = 0$ we have $u_1 = 0$. We do $x_1, x_j, y_1, y_j \to x_1 + x_j, x_j, y_1, y_j + y_1$ to make $v_2 = \ldots = v_g = 0$. Then we do $x_1, x_j, y_1, y_j \to x_1, x_j + y_1, y_1 + x_j, y_j$ to make $u_2 = \ldots = u_g = 0$. This proves a).

For b) using the new transform (∗) and $\begin{bmatrix} 0 & 1 \\ 1 & 0 \end{bmatrix} * \begin{bmatrix} 0 & 1 \\ 1 & 0 \end{bmatrix} = \begin{bmatrix} 1 & 0 \\ F_{ii} & 1 \end{bmatrix}$. We may make $y_1 = 0$ after all x's are zero except x_1 and similarly we can make $u_1 = 0$ if $v_1 = 1$. The rest is the same. $\square$

Now we can give the proof of Lemma 8.5. Now $\Gamma_\alpha = \Gamma(1,2)$, this contains $T_1,\ldots,T_g$ and hence $\begin{bmatrix} 0 & 1 \\ -1 & 0 \end{bmatrix}$. The group $\Gamma_{\alpha,A}$ consists of elements of the form $\begin{bmatrix} \alpha & \beta \\ 0 & \delta \end{bmatrix}$. The conjugate by $\begin{bmatrix} 0 & 1 \\ -1 & 0 \end{bmatrix}$ has the form $\begin{bmatrix} \alpha & 0 \\ \gamma & \delta \end{bmatrix}$. The only remaining point is that T_i^2 is contained in $\Gamma_{\alpha,A}$. The rest follows directly.

§8.4 The Relationship Between $\pi'_*\mathcal{N}_A$ and H in General

Now when $\det_L(E)$ is general the sheaf $\pi'_*\mathcal{N}_A$ on $\mathrm{Reas}(W)$ is locally free. Let $A \oplus B = L$ be a decomposition of L such that $\alpha|_A \equiv 1 \equiv \alpha|_B$. The canonical theory of theta function defines an isomorphism

$$\eta : \mathbb{C}\,[B[E]] \underset{\mathbb{C}}{\bigotimes} \mathcal{O}_{\mathrm{Reas}(W)} \longrightarrow \pi'_*\mathcal{N}_A \ .$$

We may multiply $\eta^{\otimes 2}$ by the section $\frac{1}{C}dz_1 \wedge \ldots \wedge dz_g$ of H to get an isomorphism

$$M : \mathbb{C}\,[B[E]]^{\otimes 2} \underset{\mathbb{C}}{\bigotimes} \mathcal{O}_{\mathrm{Reas}(W)} \longrightarrow (\pi'_*\mathcal{N}_A)^{\otimes 2} \otimes \mathcal{H} \ .$$

By the previous reasoning with heights and length of theta function we know that M is an isometry where the inner product on the first sheaf in induced by the natural one on $\mathbb{C}[B[E]]$ for a definite choice of the constant C.

We want to understand the action of Γ_α on the second sheaf in terms of M. We will prove

Theorem 8.11. *There is a homomorphism $\psi : \Gamma_\gamma \to U$ where U is the unitary group of $\mathbb{C}[B[E]]^{\otimes 2}$ such that the action Γ_γ on $(\pi'_*\mathcal{N}_A)^{\otimes 2} \otimes \mathcal{H}$ corresponds to the Γ_α-action on $\mathbb{C}[B[E]]^{\otimes 2} \otimes_{\mathbb{C}} \mathcal{O}_{\mathrm{Reas}(W)}$ via $\psi^{-1}\otimes$ (pull-back of functions).*

Remark. It can be shown that Image of ψ is finite.

We will first prove

Lemma 8.12. *Let U be a finite dimensional Hilbert space. Let $\| \ \|$ be the natural extension of its norm to a metric on $U \otimes_{\mathbb{C}} \mathcal{O}_X$ where X is a connected complex manifold. The global sections of $U \otimes_{\mathbb{C}} \mathcal{O}_X$ with constant length are those of the form $u \otimes 1$ where u is a vector in U.*

Proof. Let $U = \mathbf{C}^n$ with the usual norm. A section of $U \otimes_{\mathbf{C}} \mathcal{O}_X$ is a vector $(f_1(x), \ldots, f_n(x))$ of holomorphic functions of x in X. We need to show that $g(x) = \sum_j |f_j(x)|^2$ is constant if and only if the f_i's are constant. We will show that if $g(x)$ has a local maximum at x_0 then the f_i's are constants.

Choose local coordinates at x_0 so that $f_j(x)$ is given by a power series $\sum_{(n)} a_{(n),j} z^{(n)}$ which converges when $|z_j| \leq 1$ for all j. Then

$$\sum_j a_{(0),j}\overline{a_{(0),j}} \geq \sum_{(m)(m)} \sum_j a_{(m),j}\overline{a_{(m),j}} z^n \cdot \bar{z}^m \quad \text{for} \quad |z_j| = 1 .$$

Averaging over $U(1)^n$ we get

$$\sum_j a_{(0),j}\overline{a_{(0),j}} = \sum_n \sum_j a_{(n),j}\overline{a_{(n),j}} .$$

Hence $a_{(n),j} = 0$ if $(n) \neq (0)$. Therefore each $f_j(x)$ is locally constant and hence globally because X is connected. $\qquad\square$

Corollary 8.13. *Any holomorphic isometry of $U \otimes_{\mathbf{C}} \mathcal{O}_X$ has the form $\psi \otimes 1$ where ψ is an isometry of U.*

Proof. The isometry preserves the length and holomorphicness of sections. Hence it induces an isometry $\psi \otimes 1$ of $U \otimes 1$ which determines it by linearity. $\quad\square$

Theorem follows formally from this corollary.

§ 8.5 Projective Embedding of Some Moduli Spaces

Let W be a real symmetric space of dimension $2g$ with skew-form E. Let $L \subset W$ be a lattice with elementary divisors $e_1|\ldots|e_g$. Let $A \otimes B = L$ be a decomposition of L. Then as usual we have $L^{\widehat{}} = A^{\widehat{}} \oplus B^{\widehat{}}$. Let Γ be the kernel of the natural mapping $\mathrm{Sym}_{\mathbf{Z}}(L) \to \mathrm{Aut}(L^{\widehat{}}/L)$. Here $L^{\widehat{}}/L = A^{\widehat{}}/A \oplus B^{\widehat{}}/B \equiv A(E) \oplus B(E)$. Let $\alpha : L \to \{\pm 1\}$ be the unique solution of the equation $\alpha(l_1 + l_2) = \alpha(l_1) \circ \alpha(l_2)(-1)^{E(l_1, l_2)}$ and $\alpha|A \equiv 1 \equiv \alpha|B$.

Then we have the invertible sheaf $\mathcal{N}_A$ on B/L. We have the universal theory of theta functions $\eta : \mathbf{C}[B[E]] \to \pi'_* \mathcal{N}_A$. Then η is induced by a linear transform $\bar{\eta} : \mathbf{C}[B[E]] \to \Gamma(B/L, \mathcal{N}_A)$.

Lemma 8.14. *If $e_1 \geq 2$ the image of $\bar{\eta}$ has no base-points.*

Proof. For each J in $\mathrm{Reas}(W)$, restriction $\bar{\eta} : \mathbf{C}[B] \xrightarrow{\approx} \Gamma(X(J), \mathcal{N}_{A,J})$ is an isomorphism. The assumption means that $\mathcal{N}_{A,J}$ is an e_1-power. Thus this result follows from Lemma 2.7. $\qquad\square$

Let $\mathbb{P}^n$ be the projective space of hyperplane in $\mathbb{C}[B]$. Thus δ_b for b in $B[E]$ are a basis of the linear forms on $\mathbb{P}^n$. Also by the lemma we have an analytic mapping $\psi : B/L \to \mathbb{P}^n$ defined by $\bar{\eta}$ if $e_1 \geq 2$.

By the work of the last chapter it follows that ψ is Γ-equivalent for the trivial action on $\mathbb{P}^n$. If $e_1 \geq 3$ then $L^{\wedge}/L$ contains the e_1-torsion in $X(J)$ for all J. Thus by Theorem 9.7 Γ does not have a fixed point in $\mathrm{Reas}(W)$. Hence $\pi'' : \Gamma \backslash B/L \to \Gamma \backslash \mathrm{Reas}(W)$ is a family of abelian varieties parameterized by the manifold $\Gamma \backslash \mathrm{Reas}(W)$. Now we have an induced analytic mapping $\overline{\psi} : \Gamma \backslash B/L \to \mathbb{P}^n$. Composing with the zero section we have another $\overline{\psi} : \Gamma \backslash \mathrm{Reas}(W) \to \mathbb{P}^n$.

Theorem 8.15. *(Mumford) If e_1 is an even integer ≥ 4 then $\overline{\overline{\psi}}$ is an embedding.*

Proof. $\overline{\overline{\psi}}$ sends a polarized abelian variety $(X(J), H_J)$ with an identification of $A(E) \oplus B(E)$ with $K(\mathscr{L}(1, H_J))$ to the point $(\delta_b|_{\sigma b \in B[E]})$ in $\mathbb{P}^n$. By Theorem 6.12 and 6.13 these null-theta werte determine the equation of $X(J)$ in $\mathbb{P}^n$ and the zero of the image. The rest of the marking $A(E) \oplus B(E)$ can be found by applying the automorphism $\bar{k}$ of $\mathbb{P}^n$ to the image of zero point where k is a point of $A(E) \oplus B(E)$. This shows that $\overline{\overline{\psi}}$ is injective. Doing the same argument over $\mathrm{Spec}(\mathbb{C}[\epsilon]/\epsilon^2)$ we see that it is also infinitesimally injective. $\square$

Let Δ_b be coordinates in another copy of $\mathbb{P}^n$. The universal family of abelian varieties with marked structure is given by $X(a+b, a-b, \chi)|_0 \widetilde{X}(c+d, c-d, \chi) - X(c+b, a-b, \chi)|_0 \widetilde{X}(c+a, c-a, \chi) = 0$ where $\widetilde{X}(e, f, \chi) = \sum_{j \in J} \chi(j) \Delta_{e+j} \otimes \Delta_{f+j}$ in the usual situation. As $(\delta_b|_0)$ satisfies these equations, we have the (Mumford-Riemann) $X(a+b, a-b, \chi)|_0 X(c+d, c-d, \chi|_0 - X(c+d, c-b, \chi)_0 X(c+a, c-a, \chi) = 0$. These equations plus $\delta_b|_0 = \delta_{-b}|_0$ define an algebraic subscheme M.-R. of $\mathbb{P}^n$ containing the image of $\overline{\overline{\psi}}$. One can algebraically prove that image of $\overline{\overline{\psi}}$ is an open subscheme of M.-R. and has Zariski closed completement. This is discussed in [4] and [2]. In Igusa's book [1] the boundary of the moduli is studied analytically.

As a final result about moduli we will sketch.

Theorem 8.16. *$\overline{\psi}$ is an unramified 2^{2g}-degree covering of its image.*

Proof. Let x be a point on a marked abelian variety X. Then by Exercise 1 of Section 6.3 we know that $\overline{\psi}(x) = (\delta_b|_x)$ defines the image $\overline{X}$ of X in $\mathbb{P}^n$ but we don't know the zero point of $\overline{X}$. The zero point y must satisfy $(\delta_b|_e) = (\delta_{-b}|_e)$ which means that $\delta_b|_e \sim \delta_b|_{-e}$ by the inverse formula. By this means that

$e = -e$ as a point X. Hence the origin of $\overline{X}$ is only known upto 2-torsion in X. Now if $\overline{\psi}(x_1) = \overline{\psi}(x_2)$ then we have an isomorphism of affine space $\psi : X_1 \to X_2$ which takes 0 to a two torsion point such that $\psi|_{X_2} \cdot \phi = \psi|_{X_1}$. The only idea is $\#X_2 = 2^{2\dim X}$. $\qquad\qquad\square$

Chapter 9. Mappings to Abelian Varieties

§ 9.1 Integration

Let $X = V/L$ be a complex torus and let S be a smooth connected variety. Let s_0 be a fixed point of S. Let $f : S \to V/L$ be an analytic mapping. Then we have the pull-back mapping of differentials $f^* : V^* = \Gamma(X, \Omega_X) \to \Gamma(S, \Omega_S)$.

Lemma 9.1. *a)* f *is determined by* $f(s_0)$ *and* $f^* : V^* \to \Gamma(S, \Omega_S)$.

b) Conversely given a point x *of* X *and a homomorphism* $\alpha : V^* \to \Gamma(S, \Omega_S)$ *these have the form* $f(s_0)$ *and* f^* *if and only if for a closed path* σ *in* S *the element* $\int_\sigma \alpha(\lambda)$ *for* λ *in* V^* *is evaluation at an element* l *of* L.

Proof. For a) let $(\widetilde{S}, 0)$ be the universal covering space of (S, s_0) with covering mapping π. Then $f \circ \pi : \widetilde{S} \to X$ lifts to an analytic mapping $f\widetilde{\circ}\pi : \widetilde{S} \to V$ where $f\widetilde{\circ}\pi = v_0$ where $f(s_0) = v_0 + L$. Clearly $\langle f\widetilde{\circ}\pi(\tilde{s}), \lambda \rangle = \int_0^{\tilde{s}} \pi^* f^*(\lambda)$ for all λ in V^*. Thus $f\widetilde{\circ}\pi$ is determined by f^* and v_0. Hence f is determined by f^* and $f(s_0)$.

For b) we define $g : \widetilde{S} \to V$ by $\langle g(\tilde{s}), \lambda \rangle = \int_0^{\tilde{s}} \pi^* \alpha(\lambda) + k$ where $k + L = x$. Thus g is analytic and we want the composition to $\widetilde{S} \to V \to V/L$ to factor through $\pi : \widetilde{S} \to S$. This is exactly the given condition. So b) follows easily. $\quad\square$

A special thing happens when X is abelian.

Proposition 9.2. *Let* f *be a meromorphic mapping of a smooth variety into an abelian variety* $X = V/L$. *Then* f *extends to an analytic mapping.*

Proof. In this case $f^* : V^* \to$ meromorphic differentials on S. We need to see that a differential in $f^*(V^*)$ has no pole along any particular divisor D on S. By Theorem 2.11 we have a projective embedding $i : X \subset \mathbb{P}^n$ where $i \circ f(s) = (\sigma_0(s), \ldots, \sigma_n(s))$ with meromorphic functions. Let $\sigma_i(s)$ have the worst pole of these functions on D. Then $i \circ f(s) = (\sigma_0(s)/\sigma_i(s), \ldots, \sigma_n(s)/\sigma_i(s))$ is regular and non-zero generally along D. Thus f is analytic generally along D and, hence, $f^*(V^*)$ consists of differentials which are regular along D. $\quad\square$

Exercise 1. Let $f : X \to Y$ be an analytic mapping between two complex tori. Then $f(x) = f(0) + g(x)$ where $g(x)$ is a homomorphism $g : X \to Y$.

§ 9.2 Complete Reducibility of Abelian Varieties

Let X be a complex torus and let Y be a subtorus. A complement to Y is a subtorus Z such that $X = Y + Z$ and the intersection $Y \cap Z$ is finite. In this case we have an isogeny $Y \oplus Z \to X$ and $\dim Y + \dim Z = \dim X$.

Proposition 9.3. *If X is an abelian variety then any subabelian variety Y has a complement.*

Proof. Let $X = V/L$ and $Y = W/M$ where W is a subspace of V and $M = W \cap L$. Let H be a polarization of X. Let $U = \{v \in V | H(v, w) = 0$ for all w in $W\}$. This is a complex subspace of V and $W \oplus U = V$ as H is positive definite. As $U = \{v \in V | E(v, lm) = 0$ for all m in $M\}$ where $E = \operatorname{Im} H$. It follows that $N = U \cap L$ is a lattice in U. The subtorus U/N is complementary to Y by construction. $\qquad\square$

A complex torus is called simple if its only proper subtori are zero. With this definition we have

Corollary 9.4. *An abelian varieties X contains a finite number of simple subabelian varieties $X_1, \ldots, X_r$ such that the natural homomorphism $X_1 \oplus \ldots \oplus X_r \to X$ is an isogeny.*

Proof. If X is simple then there is no problem. Otherwise a minimal non-zero subabelian variety X_1 of X is simple. By Proposition 9.3 X_1 has a complement Y. As $\dim Y < \dim X$ we may assume that Y has a finite number of simple subtori $X_2, \ldots, X_r$ such that $X_2 \oplus \ldots \oplus X_r$. Therefore the second statement of the corollary is true for these X_i's. $\qquad\square$

Isogeny between two complex tori is an equivalence relation. Clearly it is transitive and contains the identity. The following lemma shows that it is reflexive.

Lemma 9.5. *Let $f : X \to Y$ be an isogeny of complex tori. Then there is an isogeny $g : Y \to X$ such that $f \circ g = (\deg f) \cdot 1_Y$.*

Proof. We may assume that $f : V/L \to V/M$ is induced by the identity on V. Then $\#(M/L) = \deg f$. Hence $(\deg f)M \subset L$. Thus we have isogenies $V/M \xrightarrow{\deg f} V/(\deg f)M \xrightarrow{g} V/L \xrightarrow{f} V/M$ and our lemma is obvious. $\square$

For X a complex torus V/L, the ring of endomorphism $\mathrm{End}(X)$ is a subring of $\mathrm{End}_{\mathbf{Z}}(L)$. Hence $\mathrm{End}(X)$ is a finitely generated $\mathbf{Z}$-module. Let $\mathrm{End}^0(X)$ be the $\mathbb{Q}$-algebra $\mathrm{End}(X) \otimes_{\mathbf{Z}} \mathbb{Q}$. As $\mathrm{End}^0(X) = \{P \in \mathrm{Hom}_{\mathbb{C}}(V, V) | P(L \otimes_{\mathbf{Z}} \mathbb{Q}) \subset L \otimes_{\mathbf{Z}} \mathbb{Q}\}$, we see that $\mathrm{End}^0(X)$ is an isogeny invariant and an isogeny has an inverse in $\mathrm{End}^0(X)$.

Now let X be an abelian variety. By Corollary 9.4 we may find non-isogenous simple abelian varieties $Y_1, \ldots, Y_g$ such that X is isogenous to $\bigoplus_{1 \leq j \leq g} Y_j^{n_j}$ for some positive integers $n_1, \ldots, n_g$. Now $\mathrm{End}^0(X) = \bigotimes_{1 \leq j \leq g} M(n_j, \mathrm{End}^0(Y_j))$ where $M(r, A)$ is the matrix algebra of degree r with coefficients in A. On the other hand $\mathrm{End}^0(Y_j)$ is a division algebra because Y_j is simple. Thus the problem of determining $\mathrm{End}^0(X)$ is reduced to the case where Y is simple. I refer the reader to Mumford's book [3] for an introduction to the classification of all such division algebras.

The key idea is the Rosatti involution $\psi \mapsto \phi_{\mathscr{L}}^{-1} \circ \psi^{\widehat{}} \circ \phi_{\mathscr{L}} = \psi'$. The main fact is the positivity of the form $\mathrm{Tr}(\psi \cdot \psi')$ on $\mathrm{End}^0(X)$.

§9.3 The Characteristic Polynomial of an Endomorphism

Let f be an endomorphism of a complex torus $X = V/L$. Then f induces an endomorphism $\tilde{f}$ of L. The characteristic polynomial of $\tilde{f}$ is called the characteristic polynomial $P_f(t)$ of f. As $\tilde{f}$ is given by an integral matrix, the coefficients of $P_f(t)$ are integers. Furthermore f satisfies the characteristic equation $P_f(f) = 0$ as an endomorphism of X.

Let g be the dimension of X. Then $P_f(t)$ has degree $2g$ and is given by

$$P_f(t) = t^{2g} - \mathrm{Tr}\, \tilde{f} t^{2g-1} + \ldots + \mathrm{Tr}(\wedge^{2g} \tilde{f}) \ .$$

The constant term $\det \tilde{f}$ is just the scalar by which $\wedge^{2g} \tilde{f}$ acts on $\wedge^{2g} L \simeq H^{2g} H^1(X, \mathbf{Z}) \simeq H^{2g}(X, \mathbf{Z})$. It is elementary that $\det \tilde{f}$ is the degree of f. Therefore we have proven

Lemma 9.6. *a)* $P_f(t) = \deg(t \cdot 1_X - f)$ *for all integers t.*

b) The characteristic roots of $\tilde{f}$ are algebraic integers.

As an application we will prove.

Proposition 9.7. *Let $X = V/L$ be an abelian variety with polarization H. Let $\operatorname{Aut}(X, H)$ be the group of automorphisms of X which preserve H. Consider the restriction $\varphi_n : \operatorname{Aut}(X, H) \to \operatorname{Aut}(X_n)$ for some integer n.*

a) If $n \geq 3$, φ_n is injective.

b) $\operatorname{Kernel}(\varphi_2)$ is a finite group of idempotents.

Proof. We will first prove that $\operatorname{Aut}(X, H)$ is a finite group. An automorphism of (X, H) is an automorphism of V which preserve both H and L. Thus $\operatorname{Aut}(X, H) = $ unitary group of $H \cap \operatorname{GL}(L)$. As the intersection of a compact group with a descrete group is finite, $\operatorname{Aut}(X, H)$ is finite.

Let φ be an element of $\operatorname{Kernel}(\varphi_n)$. As φ has finite order, $\tilde{\varphi}$ is semi-simple and has roots of unity as eigenvalues. Clearly it is enough to show that the order p of an eigenvalue η cannot be an odd prime or either two if $n \geq 3$ or four if $n = 2$.

Now $\varphi - 1$ is zero on X_n. Thus $\varphi - 1 = n\alpha$ where α is an endomorphism of α. The eigenvalues k of α are algebraic integers as satisfy $\eta - 1 = nk$. Thus as α is integral, $n^{\deg_\mathbb{Q}(\eta)}$ divides $\operatorname{Norm}_\mathbb{Q}(\eta - 1)$. This is impossible. Explicitly if $p = 4$ and $n = 2$, $\operatorname{Norm}_\mathbb{Q}(\pm\sqrt{-1} - 1) = 2$ is not divisible by 2^2. If p is an odd prime or $n \geq 3$, $\operatorname{Norm}_\mathbb{Q}(\eta - 1) = \prod_{1 \leq i \leq p-1}(\eta^i - 1) = \frac{(1 - X^p)}{1 - X}\big|_{X=1} = -\frac{d}{dx}(1 - Xp)\big|_{X=1} = p$ is not divisible by n^{p-1}. $\qquad\square$

§ 9.4 The Gauss Mapping

Let S be a subvariety of dimension s of a complex torus $X = V/L$. By translation we identify the tangent space of X at any point with V. Let s be a smooth point of S, then we have the s-dimensional subspace $T_s S$ of V where $T_s S$ is the tangent space of S at s. This defines a holomorphic mapping

$$G : S_{\text{smooth}} \longrightarrow \operatorname{Grass}_s V$$

into the Grassmann manifold of s-planes in V. The objective of this section is to prove

Proposition 9.10. *We have two mutually exclusive possibilities;*

a) G is generically finite-to-one, and

b) $S = \pi^{-1} S'$ where S' is a subvariety of an abelian variety X' and $\pi : X \to X'$ is a non-trivial surjective homomorphism.

Proof. First let Y be an irreducible closed analytic subspace of X. The complex torus $\langle Y \rangle$ spanned is $\sum_{0 \leq n \leq \infty}(Y - Y)$ consisting of all difference of the sum of n-terms from Y. To check that this is actually a complex torus, first note that $\sum_{1 \leq n \leq i}(Y - Y)$ is a closed irreducible analytic subset of X as it is the image of an irreducible compact variety Y^{2i} by a morphism. Thus for dimension the above increasing union terminates in an equality $\langle Y \rangle = \sum_{0 \leq i \leq n}(Y - Y)$ for $n \gg 0$. Clearly $\langle Y \rangle$ is a subgroup and hence $\langle Y \rangle$ is complex torus. By a variant of Theorem 9.1 the tangent space of $\langle Y \rangle$ is the span of the tangent space Y and y for y in a open subset of Y.

Let s be a point of S_{smooth}. Let Y_s be the irreducible component of $(s, G(s))$ in the closure of the graph of G. Thus Y_s is analytic. We will assume that G has maximal rank at s. Thus for these s the tangent space of $\langle Y_s \rangle$ depends analytically on s. We may furthermore restrict s so that the dimension of the subspace is maximal and hence these spaces form a vector bundle. Thus $\langle Y_s \rangle$ is a smooth family of subabelian varieties. If we consider how its tangent space meets the lattice L, we conclude that this family $\langle Y_s \rangle$ is a constant variety Y.

Next let $X' = X/Y$ and $S' = \pi(S)$ which is analytic by Grauert theorem again. We want to show that $S = \pi^{-1}(S')$. To do this it is enough to show that for s in an open subset S the kernel K of differential of π at s contains the tangent space N of Y; i.e. $N \subset K$. Now for good s' N depends only on $\pi(s)$. Thus $N \supset \sum_{\pi(s')=\pi(s)}$ tangent space of Y_s at $s' = K$. $\qquad\square$

Corollary 9.11. *If D is an irreducible ample divisor on an abelian variety X, the Gauss mapping $D_{\text{smooth}} \to \{Hyperplanes\ in\ V\}$ is generically finite-to-one.*

Proof. As D is ample it can not be the pull-back of a divisor by a non-trivial surjection. $\qquad\square$

It would be interesting if this result is also true in characteristic p (even in the principally polarizied case).

Chapter 10. The Linear System $|2D|$

§ 10.1 When $|D|$ Has No Fixed Components

Let D be an ample divisor on an abelian variety X. Recall by Lemma 2.7 the complete linear system $|2D|$ has no base points. Let $\varphi : X \to \mathbb{P}^n$ be the morphism defined by $|2D|$.

In this section we will study the case where $|D|$ has no fixed components; i.e. $\mathrm{cod}(\bigcap_{E \in |D|} E) > 1$. In other words for any effective divisor F of X, the space $\Gamma(X, \mathcal{O}_X(D - F))$ is a proper subspace of $\Gamma(X, \mathcal{O}_X(D))$.

We recall from Section 2.4 that a general divisor E in $|D|$ is reduced and has the property that if $E + x = E$ then $x = 0$.

Theorem 10.1. *If $|D|$ has no fixed components, then $\varphi : X \to \mathbb{P}^n$ is an embedding.*

Proof. We will first show that φ is injective. Let x and y be two points of X such that $\varphi(x) = \varphi(y)$. We want to prove that $x = y$. By assumption $x \in E \Leftrightarrow y \in E$ for all E in $|2D|$. Let G be a divisor in $|D|$ such that G is reduced. Let H be another divisor in $|D|$ such that H is reduced, no component of H is a component of $(-G) + x + y$ and $"H = H + z" \Rightarrow" z = 0"$. We know by the theorem of the square 2.7 that for any point z of X the divisor $(G + z) + (H - z)$ is contained in $|2D|$. Thus $x \in A(z) \Leftrightarrow y \in A(z)$. This means that $z \in ((-G) + y) + (H - y) \Leftrightarrow z \in ((-G) + x) + (H - x)$. Thus $((-G) + y) + (H - y) = ((-G) + x) + (H - x)$ set-theorically. So $H \subset ((-G) + x + y) + (H - x + y)$. Hence $H \subset H - x + y$ but these reduced divisors are equivalent. Therefore $H = H - x + y$. Hence $0 = -x + y$ or rather $x = y$.

It remains to show that φ is infinitesimally injective. Let $X = V/L$. Let v be a vector in V such that $(T_z\varphi)_z(v) = 0$ where $T_z\varphi$ is the linear approximation of φ at some point z of X. We want to see that $v = 0$. By assumption if z is contained in a divisor E in $|2D|$ then E is tangent to v at z. Let G still be a reduced divisor in $|D|$. Let H now be a reduced divisor in $|D|$ such that no component of $(-G) + 2z$ is contained in H. We know that v is tangent to $(G + k) + (H - k)$ at z if $z \in (G + k) \Leftrightarrow k \in (-G) + z$. If for general k in any component C of $(-G)+z$, $H-k$ does not contain z. Otherwise k in C would be contained $H-z$; i.e. a component of $(-G)+2z$ is contained in H which is forbidden. Thus v is tangent to G at all its points and we are done as in the proof of Theorem 2.11. $\qquad\square$

§ 10.2 Projective Normality of $|2D|$

Let $\mathscr{L}$ be the invertible sheaf $\mathcal{O}_X(2D)$. Now $\mathscr{L}$ is normally generated if the graded ring $\bigoplus_{n\in\mathbf{Z}} \Gamma(X,\mathscr{L}^{\otimes n})$ is generated by $\Gamma(X,\mathscr{L})$. This is equivalent to $\varphi : X \to \mathbb{P}^n$ being an embedding and the cone of X in A^{n+1} being normal; i.e. $X \subset \mathbb{P}^n$ is projectively normal.

Assume that $\mathcal{O}_X(D) = \mathcal{M}$ is excellent for some decomposition $L = A \oplus B$ where $X = V/L$.

Theorem 10.2. *$\mathscr{L}$ is normally generated if and only if no point of $(2)^{-1}K(\mathcal{M})$ is a base-point of $|D|$.*

Proof. If $n \geq 2$ the multiplication $\Gamma(X,\mathscr{L}) \otimes \Gamma(X,\mathscr{L}^{\otimes n}) \to \Gamma(X,\mathscr{L}^{\otimes n+1})$ is always surjective by Theorem 6.8 c). Thus $\mathscr{L}$ is normally generated if and only if the multiplication $M : \Gamma(X,\mathscr{L}) \otimes \Gamma(X,\mathscr{L}) \to \Gamma(X,\mathscr{L}^{\otimes 2})$ is surjective. The other part of the theorem is equivalent to the statement that o is not a base-point of $\Gamma(X,\mathcal{M} \otimes P_\alpha)$ for any two-torsion point α of the dual variety $X^{\widehat{\ }}$. This follows because the $\mathcal{M} \otimes P_\alpha$ are exactly the sheaves $T_k^*\mathcal{M}$ where k runs through $(2)^{-1}K(\mathcal{M})$. Thus we want to show that M is surjective if and only if for all α in $(X^{\widehat{\ }})_2$ there is a section φ_α of $\mathcal{M}\otimes P_\alpha$ which does not vanish at zero.

Recall from Theorem 6.7 that

$$M(X(b_1' + b_2',\, b_1' - b_2',\chi)) = Y(b_2',\chi)|_0 \cdot Y(b_1',\chi)$$

for all character χ of $B(\mathscr{L})_2$ and b_1' and b_2' in $B(\mathscr{L}^2)$ which are congruent modulo $B(\mathscr{L})$. Recalling from Section 6.1 that the Y's and X's are basis of their respective spaces after we account for the obvious $B(\mathscr{L})_2$ redundances.

Fix b_1' in $B(\mathscr{L}^2)$ then the image of M intersected with $\mathbb{C}Y(b_1',\chi)$ is $Y(b_1',\chi)|_0 \cdot Y(b_1',\chi)$ where $b_1' \equiv b_2'(B(\mathscr{L}))$. Thus M is surjective if and only if for all χ and b in $B(\mathscr{L}^2)$ $Y(b',\chi)|_0 \neq \mathbf{Z}$ for some $b' \equiv b(B(\mathscr{L}))$. Now by definition these $Y(b',\chi)$ are the span of the sections of $\mathscr{L}^{\otimes 2} = \mathcal{M}^{\otimes 4} = 2^{\mathcal{M}}$ of $2^*\mathcal{M}$ with X_2-eigenvalue $(a,b) \mapsto (e_{\mathscr{L}^2}(a,b')^{-1},\chi(b))$. Therefore they are the pull-back by 2 of a basis of $\mathcal{M} \otimes P_\alpha$ where α is the two-torsion point of $X^{\widehat{\ }}$ corresponding to the eigenvalue. As all of $X_2^{\widehat{\ }}$ occurs this way we have shown that M is surjective if and only if for each α there is a section φ_α of $\mathcal{M} \otimes P_\alpha$ which does not vanish at zero. $\qquad\square$

§ 10.3 The Factorization Theorem

Let E and F be effective divisors on an abelian variety X such that $E + F$ is ample. We have the obvious inclusion $r : \Gamma(X, \mathcal{O}_X(E)) \hookrightarrow \Gamma(X, \mathcal{O}_X(E + F))$.

Theorem 10.3. *If r is an isomorphism, then X is isomorphic to product $X_E \times X_F$ and under this isomorphism E and F correspond to divisors of the form $E' \times X_F$ and $X_E \times F'$ where E' and F' are ample divisors of X_E and X_F respectively and F' gives a principal polarization of X_F.*

Proof. By Chap. 2 if we let Y_E and Y_F be the connected components of $K(\mathcal{O}_X(E))$ and $K(\mathcal{O}_X(F))$ then $E = s_E^{-1}(E')$ and $F = s_F^{-1}(F')$ where E' is an ample divisor on $X_E = X/Y_E$ and $s_E : X \to X_E$ is the projection and similar with F. Thus we have the sum $s = (s_E, s_F) : X \to X_E \times X_F$. We intend to show that s is an isomorphism and $s^{-1}(E' \times X_F) = E$ and $s^{-1}(X_E \times F') = F$.

Now $\mathcal{O}_X(E)$ is trivial on X_E and similarly with F. Thus the ample sheaf $\mathcal{O}_X(E+F)$ is trivial on the intersection $X_E \cap X_F = \mathrm{Ker}\, s$. Hence the kernel of s is finite. By the Riemann-Roch theorem $\frac{1}{x!}(E+F)^x = \dim \Gamma(\mathcal{O}_X(E+F))$ where $x = \dim X$. Furthermore $\frac{1}{x_E!}(E)^{x_E} = \dim \Gamma(\mathcal{O}_{X_E}(E)) = \dim \Gamma(\mathcal{O}_X(E))$ where $x_E = \dim X_E$ and similarly with F. Therefore $\frac{1}{x!}(E + F)^x = \frac{1}{x_E!}(E')^{x_E} \equiv 1$ and $\frac{1}{x_F!}(F')^{x_F} \geq 1$. By the binomial theorem $a = \frac{1}{x!}(E+F)^x = \sum_{e+f=x} \frac{1}{e!}E^e \cdot \frac{1}{f!}F^f$. As E and F are effective each one of these summand ≥ 0. We will estimate some. Consider $b = \frac{1}{x_E!}E^{x_E} \cdot \frac{1}{f!}F^f$ where $x_E + f = x$. Now $\frac{1}{x_E!}E^{x_E} = s_E^{-1}(\frac{1}{x_E!}(E')^{x_E})$ is equivalent to a coset of $\ker(s_E) = s_E^{-1}(0) = Y_E$. Thus $b = a(Y_E \cdot \frac{F^f}{f!})$. Now by the projection formula $Y_E \cdot F^f = s_{F*}(Y_E) \cdot F^f$ but $s_{F*}|Y_E$ is a finite morphism onto its image say Z and the degree $= \#(Y_E \cap Y_F)$. Thus $Y_E \cdot \frac{F^f}{f!} = \#(Y_E \cap Y_F)^f$ where $f = \dim Z$. As F' is ample $(\frac{F'|_Z}{f!})$ is a positive integer k. Therefore $a \geq b \geq a \cdot \#(Y_E \cap Y_F) \cdot k$. Hence $\ker s = Y_E \cap Y_F = 0$ and $k = 1$ and b is the only non-zero summand of a. Reversing the role of E and F we see that $\frac{1}{e!}E^e \frac{1}{x_F!}F^{x_F}$ is non-zero where $e + x_F = x$. Thus $e = x_E$ and $f = x_F$. In particular $X_E \times X_F$ has the same dimension as X and s has degree 1. Hence s is an isomorphism. The rest is clear. $\qquad\square$

Famous special case is

Corollary 10.4. *Let D be an effective ample division of an abelian variety. If D gives a principal polarization, (X, D) is a product $\times(X_i, D_i)$ of principally polarized abelian varieties X_i with irreducible divisors D_i.*

Proof. If $D = E + F$ is reducible then r is automatically an isomorphism of lines (!). Hence $(X, E + F)$ factors as in the theorem. Repeat until the divisors are irreducible. $\qquad\Box$

Another corollary is useful for studying an ample linear system $|D|$ on an abelian variety.

Corollary 10.5. $|D| = |M| + F$ *where* $|M|$ *is a linear system without fixed components on* $X = X_M \times X_F$ *where* $F = X_M \times F'$ *and* $M = M' \times X_F$ *where* F' *gives a principal polarization of* X_F.

Proof. Let F be the greatest common divisor of $|D|$. Then let $E = M + F$ for some E in $|D|$. By construction $r : \Gamma(X, \mathcal{O}_X(M)) \to \Gamma(X, \mathcal{O}_X(E))$ is an isomorphism and Theorem 10.3 applies. $\qquad\Box$

§ 10.4 The General Case

Let D be an effective ample divisor on a abelian variety X. By the last two corollaries we may factor (X, D) as $(Y, M) \times X(X_i, P_i)$ where $|M|$ has no base components on Y and P_i is an irreducible principal polarization of Z_i. Thus by the Künneth formula $\varphi : X \to \mathbb{P}^n$ for $|2D|$ is the product of the correspondence morphism on each factor. As we understand the moving case (Y, M) by Theorem 10.1 we only need study the case when D is an irreducible principal polarization of X.

Assume that this is the case and also that $\mathcal{M} = \mathcal{O}_X(D)$ is excellent for some decomposition $L = A \oplus B$ where $X = V/L$. Now $\mathcal{M} = (-1)^* \mathcal{M}$ and (-1) acts on $\Gamma(X, \mathcal{M}^{\otimes j})$ for any j.

As $\mathcal{M}$ is principal, $K(\mathcal{M}) = 0$. Hence $K(\mathcal{M}^{\otimes j}) = X_j$ and we have the decomposition $A(\mathcal{M}^{\otimes j}) \oplus B(\mathcal{M}^{\otimes j})$ of X_j. By Chapter 5 we have the theory

$$\eta_{\mathcal{M}^j} : \mathbb{C}(B(M^{\otimes j})) \underset{\approx}{\to} \Gamma(X, \mathcal{M}^{\otimes j}) \ .$$

By the Theorem 6.4, we know the effect of $(-1)^*$ corresponds to the mapping $((-1)^* f)(b) = f(-b)$ on functions on $B(\mathcal{M}^{\otimes j})$. When $j = 2$, $B(\mathcal{M}^{\otimes 2})$ consists entirely of 2-torsion. Hence $(-1)^*$ is the identiy on $\Gamma(X, \mathcal{M}^{\otimes 2})$. In other words, every section of $\mathcal{M}^{\otimes 2}$ is even. This implies that $\varphi : X \to \mathbb{P}^n$ factors through the quotient morphism $X \to X/\{\pm 1\}$ where $X/\{\pm 1\}$ is the Wertinger (or so-called Kümmer) variety. In this section we have

Theorem 10.6. *The natural morphism* $\varphi' : X/\{\pm 1\} \to \mathbb{P}^n$ *is an embedding.*

Proof. We will first show that φ' is injective. Let x and y be two points of X such that $\varphi'(x) = \varphi'(y)$. We want to show that $x = \pm y$. Let $A(z) = (D + z) + (D - z)$ be the divisor in $|2D|$ as usual (see the proof of Theorem 10.1). Then $x \in A(z) \Leftrightarrow y \in A(z)$ or, rather, $z \in ((-D) + x) + (D - x) \Leftrightarrow z \in ((-D) + y) + (D - y)$. So $((-D) + x + (D - x)) = ((-D) + y) + (D - y)$. There are two cases: a) $(-D) + x = D - y$ or b) $D - x = D - y$. In the second case $D = D + (x - y)$ or $x = y$. In the first case we have $D = -D$ as $(-1)^*$ acts trivially on $\Gamma(X, \mathcal{O}_X(D))$. So $D = D + (-y - x)$ or $x = -y$. The first step is done.

There are two kinds of points of $X/\{\pm 1\}$. There are special ones which are the image of X_2 and ordinary ones are not. In other words the special ones are the image of points fixed by the involution -1 on X. Therefore $X/(\pm 1)$ is smooth at an ordinary point and the projection $X \to X/(\pm 1)$ is an isomorphism on tangent space here. Next we will show that φ' is infinitesimally injective at an ordinary point which is the image of a point z of X. In other words if $X = V/L$ and v is a vector in V then $T_z(v) = 0$ implies that $v = 0$.

This is as easy as before. We know that v is tangent to $A(y)$ at z if z is a point of $A(y)$ (in particular if $z \in D - y \Leftrightarrow y \in D - z$). Either $z \notin D + y$ for general y in $D - z$ and we are done as usual or $z \in D + y$ if $y \in D - z$. This means that $y \in (-D) + z = D + z \Leftrightarrow y = D - z$ or $D + z = D - z$ or $D + 2z = D$ or $2z = 0$ which is impossible because z is not two torsion. Thus φ' is infinitesimally injective at ordinary points.

Let z be a two-torsion point of X. Let z' be its image in $X/\{\pm 1\}$. We need to see that $T_{z'}F'$ is injective. We first have to compute the tangent space of $X/\{\pm 1\}$ at z' in terms of data of X at z. Recall that a point of the tangent space is a derivation $Q : \mathcal{O}_{X/\{\pm 1\}, z'} \to \mathbb{C}$ which is linear with respect to valuation at z'. Now $\mathcal{O}_{X/\{\pm 1\}, z'} = (\mathcal{O}_{X,z})^{\text{even}}$ which consists of all Taylor series of functions on X near z with only terms of even degrees as z is an isolated fixed point of (-1). Thus Q is given by a homogenous quadratic differential operator on X evaluated at z.

Let $\mathcal{O}_X(D)$ be $\mathscr{L}(\alpha, H)$ for some A.-H. data (α, H). Then we have an even non-zero section $\theta(v)$ of $\mathscr{L}(\alpha, H)$. If Q is in the kernel of $T_{z'}(\varphi')$ then $Q(\theta(v + w)\theta(v - w))|_{\tilde{z}} = 0$ for all w in W such tht $\theta(\tilde{z} + w)\theta(\tilde{z} - w) = 0$ where $\tilde{z}$ in V lies over z. We want to show that this means $Q = 0$. Now $\theta(\tilde{z} - w) = 0 \Leftrightarrow \theta(-\tilde{z} + w) = 0 \Leftrightarrow \theta(\tilde{z} + w) = 0$. The first implication use the evenness and the second use the functional equation

$$\theta(-\tilde{z} + w) = \alpha_{-2\tilde{z}} e^{\pi H(-\tilde{z}+w, -2\tilde{z}) + \frac{\pi}{2} H(-2\tilde{z}, -2\tilde{z})} \theta(+\tilde{z} + w) = C(w)\theta(\tilde{z} + w)$$

as $2\tilde{z} \in L$. In coordinates $Q = \sum a_{i,j} \frac{\partial}{\partial X_i} \frac{\partial}{\partial Y_j}$ when $a_{i,j}$ is a matrix of constants. Now if $\theta(v)$ vanishes at $\tilde{z} + w$ then

$$Q(\theta(v+w)\theta(v-w))\Big|_{\bar{z}} = 2\sum a_{ij}\frac{\partial}{\partial X_i}\theta(v+w)\Big|_{\bar{z}}\frac{\partial}{\partial X_j}\theta(v-w)\Big|_{\bar{z}}$$

$$= 2\sum a_{ij}\frac{\partial}{\partial X_i}\theta(v)\Big|_{\bar{z}+w}\frac{\partial}{\partial X_j}\theta(v)\Big|_{\bar{z}-w}$$

$$= 2\sum a_{ij}\frac{\partial}{\partial X_i}\theta(v)\Big|_{\bar{z}+w}\frac{\partial}{\partial X_j}\theta(v)\Big|_{-\bar{z}+w}$$

$$= 2C(w)\sum a_{ij}\frac{\partial}{\partial X_i}\theta(v)\Big|_{\bar{z}+w}\frac{\partial}{\partial X_j}\theta(v)\Big|_{\bar{z}+w}.$$

Therefore if $Q \neq 0$ we would have a nontrivial equation satisfied by the gradient of θ where $\theta = 0$. This would mean that the image of the Gauss mapping $D_{\text{smooth}} \to \{\text{hyperplane in } V\}$ is contained in a quadric hypersurface. This is impossible by Theorem 9.11. Therefore $Q = 0$. $\square$

§ 10.5 Projective Normality of $|2D|$ on $X/\{\pm 1\}$

Let $\mathscr{M}$ be an invertible sheaf on an abelian variety $X = V/L$. We will assume that $\mathscr{L}$ is symmetric; i.e. $\mathscr{M} \approx (-1) * \mathscr{M}$. If (α, H) is Appel-Humbert data for $\mathscr{M}$ this means that $\alpha(l) = \alpha(-l)(= \alpha(l)^{-1})$ for all l in L or rather $\alpha(l) = \pm 1$ for all l in L. We want a geometric interpretation of these signs.

Consider the explicit isomorphism $i : \mathscr{M} \overset{\approx}{\Longrightarrow} (-1) * \mathscr{M}$ which corresponds to the substitution $f(x)$ to $f(-x)$ in $\mathscr{L}(\alpha, H)$. Clearly $i^2 = \text{identity}$. If x in X is a fixed point of the involution -1 of $X(x$ is 2-torsion), then i defines an involution i_x of the line $\mathscr{M}|_x$. If $i_x = -1$ we say that x is odd and otherwise x is even.

If all x in X_n are even (e.g. $\mathscr{M}$ is the square of a symmetric sheaf), then $\mathscr{M}$ is called totally symmetric. If $\mathscr{M}$ is totally symmetric then $\mathscr{M}$ descends to an invertible sheaf $M/(\pm 1)$ on $X/(\pm 1)$. In fact sections of $(\mathscr{M}/(\pm 1)^{\otimes n})$ correspond exactly to sections of $\mathscr{M}^{\otimes n}$ which are invariant under $\Gamma(X, i^{\otimes n})$.

We will assume that $\mathscr{M}$ is excellent with respect to a decomposition $A \oplus B = L$ of L and that $\Gamma(X, \mathscr{M}) = 1$. Then by Theorem 10.6 we have the very ample invertible sheaf $(\mathscr{L}/(\pm 1))^2$ on $X/(\pm 1)$ where $\mathscr{L} = \mathscr{M}^{\otimes 2}$ and we want to know when the graded ring $R = \bigoplus_n R_n = \bigoplus_{n \geq 0} \Gamma(X/(\pm 1), (\mathscr{L}/(\pm 1))^n) = \bigoplus_{n \geq 0} \Gamma(X, \mathscr{L}^n)^+$ where $+$ denotes the subspace fixed by $\Gamma(X, i^n)$ is generated by its linear term R_1. By previous work this is almost true.

Lemma 10.7. *R is generated as a $\mathbb{C}$-algebra by R_1 and R_2.*

Proof. In Section 10.4 we have noted that $\Gamma(X, \mathscr{L}) = \Gamma(X, \mathscr{L})^+$. If $m \geq 2$ then by Theorem 6.8 c) the multiplication $\Gamma(X, \mathscr{L}) \otimes \Gamma(X, \mathscr{L}^{\otimes m}) \to \Gamma(X, \mathscr{L}^{\otimes(m+1)})$ is surjective. Hence by taking invariants $R^1 \otimes R^m \to R^{m+1}$ is surjective. $\square$

The main result of this section is

Theorem 10.8. *R is generated by R_1 as a $\mathbb{C}$-algebra if and only if a non-zero section of $\mathcal{M}$ does not vanish at any even two-torsion point of X.*

Proof. By the previous lemma for the first condition we need to see when $M : R_1 \otimes R_1 \to R_2$ is surjective where $R_2 = \Gamma(X, \mathscr{L}^{\otimes 2})^+$. For the second condition as in the proof of Theorem 10.2 we need to know whether a non-zero section of $\mathcal{M} \otimes P_\alpha$ does not vanish at zero where $\mathcal{M} \otimes P_\alpha \approx T_x^* \mathcal{M}$ where x is an even two-torsion point of x. Let $x \equiv \frac{1}{2}a + \frac{1}{2}b$ where a and b are in A and B. Then as $\mathcal{M}$ is excellent, $\alpha(a + b) = (-1)^{E(a,b)}$. Thus we want to have $E(a, b)$ even. By Proposition 1.8 the character χ' of X_2 given α has the form $\chi'(\frac{1}{2}\underline{a} + \frac{1}{2}\underline{b}) = (-1)^{E(\underline{a},\underline{b})}(-1)^{E(\underline{b},\underline{a})}$.

Consider a character χ of $\frac{1}{2}B/B$, let $Y(b, \chi) = \sum_{\eta \in X_2} \chi(\eta)\delta_{\bar{b}+\eta}$ for b in $\frac{1}{4}B/B$ as usual. Then $i*(Y(b, \chi)) = \chi(2b)Y(b, \chi)$ by the obvious manipulation. Thus R_2 corresponds to the span of $Y(b, \chi)$ when $\chi(2b)$. Therefore by the same reasoning as in the proof of Theorem 10.2, the first condition is verified if and only if $Y(b_2^1, \chi)|_0 \neq 0$ for all b_2^1 in $\frac{1}{4}B/B$ such that $\chi(2b_2^1) = 1$. In this case $Y(b_2^1, \chi)|_0 \neq 0$ if and only if the non-zero section of $\mathcal{M} \otimes P_\alpha$ does not vanish at 0 where α is given by the character $\chi''(\frac{1}{2}\underline{a} + \frac{1}{2}\underline{b}) = (e_{\mathscr{L}}4(\frac{1}{2}\underline{a}, b_2^1)^{-1}, \chi(\frac{1}{2}\underline{b}))$. Now $e(\frac{1}{2}a, b_2^1)^{-1} = e^{2\pi i E(\frac{1}{2}\underline{a}, b_2^1)(4)}(-1)^{E(\underline{a}, 4b_2^1)}$ by Lemma 4.2. This corresponds to an even two-torsion point by an easy calculation if and only if $\chi(2b_2^1) = 1$. $\qquad\square$

Exercise 1. Show that $\dim R^i = \frac{1}{2}(2i)^g + \frac{2^g}{2}$.

Chapter 11. Abelian Varieties Occurring in Nature

§ 11.1 Hodge Structure

Let M be a free abelian group of rank $2g$. An elementary structure on M is a complex subspace U of $M \otimes_{\mathbf{Z}} \mathbf{C}$ such that $M \otimes_{\mathbf{Z}} \mathbf{C} = U \oplus \overline{U}$ where $\bar{\ }$ is the linear mapping induced by complex conjugation in $\mathbf{C}$. Thus $\dim_{\mathbf{C}} U = \dim_{\mathbf{C}} \overline{U} = g$. Let π_U and $\pi_{\overline{U}}$ denote projective of $M \otimes_{\mathbf{Z}} \mathbf{C}$ onto U and $\overline{U}$.

There are two complex tori associated with an elementary structure. Let $P(U)$ be $\overline{U}/\pi_{\overline{U}}(M)$ and let $A(U)$ be $\mathrm{Hom}_{\mathbf{C}}(U, \mathbf{C})/M^{\perp}$ where $M^{\perp}$ is the subgroup of $\mathrm{Hom}_{\mathbf{C}}(U, \mathbf{C})$ consisting of linear functionals λ such that $\lambda \circ \pi_U + \overline{\lambda(\bar{\pi}_{\overline{U}})} \equiv \pi(\lambda)$ has integral values on M.

Lemma 11.1. *a) Both $P(U)$ and $A(U)$ are complex tori.*

b) We have a canonical isomorphism

$$A(U) \xrightarrow{\approx} \mathrm{Pic}^0(P(U)) \, .$$

Proof. For $P(U)$ we need to see that $\pi_{\overline{U}} : M \otimes_{\mathbf{Z}} \mathbf{R} \to \overline{U}$ is injective (and hence an isomorphism). Let m be an element of the kernel. Then $m = \bar{m}$ and $m = (u, 0)$. So $m = \overline{(u, 0)} = (0, \bar{u})$. Thus $u = 0$ and $m = 0$. For $A(U)$ we need to show that $\alpha : M^{\perp} \otimes_{\mathbf{C}} \mathbf{R} \to \mathrm{Hom}_{\mathbf{C}}(U, \mathbf{C})$ is an isomorphism. The point is that π defines a real isomorphism $\pi : \mathrm{Hom}_{\mathbf{C}}(U, \mathbf{C}) \xrightarrow{\approx} \mathrm{Hom}_{\mathbf{Z}}(M, \mathbf{R})$. This follows because $\pi(\lambda)$ satisfies $\overline{\pi(\lambda)(v)} = \pi(\lambda)(v)$ and hence has the form $\mu \otimes \mathbf{C}$ for μ in $\mathrm{Hom}_{\mathbf{Z}}(M, \mathbf{R})$. The rest of a) follows easily.

For b) by § 1.4 $\mathrm{Pic}^0(P(U)) = \mathrm{Hom}_{\mathbf{R}}(\overline{U}, \mathbf{R})/\mathrm{Hom}_{\mathbf{Z}}(\pi_{\overline{U}}(M), \mathbf{Z})$ where the complex structure on the Hom is given by $iK(\bar{u}) = -K(i\bar{u})$. Let $H(K)$ be element of $\mathrm{Hom}_{\mathbf{Z}}(M, \mathbf{R})$ defined by $H(K) = K \circ \pi_{\overline{U}}|_M$. Consider $C : \mathrm{Hom}_{\mathbf{R}}(\overline{U}, \mathbf{R}) \to \mathrm{Hom}_{\mathbf{C}}(U, \mathbf{C})$ which equals $\pi^{-1}H$. Clearly C is a real linear isomorphism taking $\mathrm{Hom}_{\mathbf{Z}}(\pi_{\overline{U}}(M), \mathbf{Z})$ to $M^{\perp}$. It remains to check that $C(iK) = iC(K)$. Let $\lambda = C(K)$. By the definition we have $2\,\mathrm{Re}\,\lambda(u) = K(\bar{u})$ for all u in U. If $\tilde{\lambda} = C(iK)$ we have $2\,\mathrm{Im}\,\tilde{\lambda}(u) = +2\,\mathrm{Re}(-i\tilde{\lambda}(u)) = 2\,\mathrm{Re}(\tilde{\lambda}(-iu)) = (iK)(\overline{-iu}) = -K(i \cdot i\bar{u}) = K(\bar{u})$. Thus $\mathrm{Re}\,\lambda = \mathrm{Im}\,\tilde{\lambda}$ and hence $i\lambda = \tilde{\lambda}$. $\qquad \square$

Let V/L be a complex torus. We may canonically construct an elementary structure U on L together with an isomorphism $P(U) \cong V/L$. Just let U be the kernel $(L \otimes_{\mathbf{Z}} \mathbf{C} \to V)$. Then $\dim U = \dim V$. To check that this is an elementary structure note that $U \cap \overline{U} = \operatorname{kernel}(L \otimes_{\mathbf{Z}} \mathbb{R} \to V) = \{0\}$. Clearly the above arrow induces an isomorphism $\overline{U} \to V$ which identifies $\pi_{\overline{U}}(L)$ with L.

Continuing with these notations let E be a real skew-symmetric form on V. Then we have the natural complex bilinear extension $\widetilde{E}$ of E to $V \otimes_{\mathbb{R}} \mathbf{C} = L \otimes_{\mathbf{Z}} \mathbf{C}$.

Lemma 11.2. *The following are equivalent*
   a) $E(iv, iw) = E(v, w)$ for all v and w in V, and
   b) $\widetilde{E}|_{U \times U} \equiv 0$.

Proof. Any element of U has the form $v \otimes i - (iv) \otimes 1$ for some v in V. Now $\widetilde{E}(u \otimes i - (iv) \otimes 1, \ w \otimes i - (iw) \otimes 1) = E(v, w)i^2 - E(iv, w)i - E(v, iw)i + E(iv, iw) = (-E(v, w) + E(iv, iw)) + i(E(iv, w) + E(v, iw))$. Thus the equivalence is clear. $\qquad\square$

If E is integral on $L \times L$, then it is the Chern class of invertible sheaf $\mathscr{L}$ on V/L if and only if a) or b) is verified. In this case we want to consider when the Hermitian form H on V with imaginary part E is positive definite.

Lemma 11.3. *H is positive definite if and only if $\frac{1}{i}E(u, \bar{u})$ is positive definite.*

Proof. We just compute

$$
\begin{aligned}
E(v \otimes i &- (iv) \otimes 1, -v \otimes i - (iv) \otimes 1) \\
&= E(v, v)i^2 + E(iv, v)i - E(v, iv)i + E(iv, iv) \\
&= i(E(iv, v) - E(v, iv)) = i(E(iv, v) + E(iv, v)) \\
&= 2iE(iv, v) = 2iH(v, v) \ . \qquad\square
\end{aligned}
$$

Summing up we have proven

Proposition 11.4. *There is an equivalence between polarized abelian varieties $(V/L, H)$ and elementary structures U on L together with an integral skew-symmetric form E on L such that $\widetilde{E}|_{U \times U} \equiv 0$ and $\frac{1}{i}\widetilde{E}(u, \bar{u})$ is positive definite on U.*

The last bunch of data is called a polarized Hodge structure of weight one.

§ 11.2 The Moduli of Polarized Hodge Structure

Let E be a fixed nondegenerate abelian integral skew form on a rank $2g$ free abelian group L. Let $\mathrm{Hodge}(L, E)$ be the set of all polarized Hodge structure based on (L, E). This consists of all g dimensional complex subspaces U of $L \otimes_{\mathbf{Z}} \mathbf{C}$ such that

1) $U \oplus \overline{U} = L \otimes_{\mathbf{Z}} \mathbf{C}$;
2) $\widetilde{E}|_{U \times U} \equiv 0$, and
3) $\frac{1}{i} \widetilde{E}(u, \bar{u})$ is positive definite on U.

Actually 2) and 3) $\implies$ 1) because E is zero on $U \cap \overline{U}$ but $\frac{1}{i} \widetilde{E}(u, \bar{u})$ is positive definite. Hence $U \cap \overline{U} = 0$.

To parameterize these U's we regard them as points in the Grassmannian of g-dimensional subspace of $L \otimes_{\mathbf{Z}} \mathbf{C}$. Then condition 2) means that U is an isotropic subspace. This is a complex analytic condition. In the space of all g-dimensional isotropic subspace is a compact homogeneous space under the complex symplectic group of $\widetilde{E}$. The condition 3) is open and real analytic. Thus the natural moduli space for polarized Hodge structues is an open subset M of the space of all isotropic g-dimensional subspace of $L \otimes_{\mathbf{Z}} \mathbf{C}$.

Fortunately we won't need global coordinates to describe M because it lies in a local coordinate neighborhood. The crucial fact is as follows. Let $A \oplus B = L$ be a decomposition of L.

Lemma 11.5. *Let U be a Hodge structure on L. Thus U is the graph of a homomorphism $\varphi_U : B \otimes_{\mathbf{Z}} \mathbf{C} \to A \otimes_{\mathbf{Z}} \mathbf{C}$.*

Proof. We first see that $U \cap A \otimes_{\mathbf{Z}} \mathbf{C}$. Hence $\frac{1}{i} \widetilde{E}(w, \bar{w})$ is zero because A is isotropic. Hence by 3) w is zero. Thus the projection $U \to B \otimes_{\mathbf{Z}} \mathbf{C}$ is an isomorphism. So U is a graph. $\square$

The graph φ_U gives local coordinates in M. Let us compute an example. Assume that U comes from a polarized abelian variety $(V/L, H)$ where $E = \mathrm{Im}\, H$. Let $a_1, \ldots, a_g, b_1, \ldots, b_g$ be a canonical base of L with elementary divisors $e_1 | e_2 \ldots | e_g$. Let $b'_j = b_j \otimes \frac{1}{e_j}$. Then φ_U is given by the matrix σ such that $\varphi_U(b'_j) = \sum a_k \otimes \sigma^k_j$ where $(\varphi_U(b'_j), b'_j)$ is in U, i.e. $b'_j + \sum_k a_k \otimes \sigma^k_j = 0$. Thus our matrix σ is just the usual τ attached $(V/L, H)$ with respect to $a_1, \ldots, a_g, b'_1, \ldots, b'_g$. Thus M is biholomorphic to $\mathrm{Reas}(L \otimes_{\mathbf{Z}} \mathbb{R})$ as the coordinates are the same.

From the point of view of Hodge structures these coordinates are unnatural because they are taken with respect to coordinates around the isotropic subspace $A \otimes_{\mathbf{Z}} \mathbf{C}$ which is not a Hodge structure. The obvious question is, "What happens if we take coordinates around a fixed Hodge structre U_0?".

Consider the direct sum $U_0 \oplus \overline{U}_0 = L \otimes_{\mathbb{Z}} \mathbb{C}$.

Lemma 11.6. *If U is any Hodge structure, $U \cap \overline{U}_0 = \{0\}$.*

Proof. $\frac{1}{i}E(u, \bar{u})$ is positive definite on U but negative definite (why?) on $\overline{U}_0$. Thus $U \cap \overline{U}_0 = \{0\}$. $\qquad\qquad\square$

Thus we may write U as the graph of a homomorphism $\phi_U : U_0 \to \overline{U}_0$. We want to compute the matrix representing ϕ_U for a basis in this space for a particular U_0. Let U_0 be the Hodge structure with $\sigma = iI_g$. Thus $c_k = b'_j + a_j \otimes i$ are a basis for U_0 and $d_k = \bar{c}_k = b'_j - a_j \otimes i$ are a basis of $\overline{U}_0$. Let $\phi_U(c_k) = \sum_l w^l_k d_k$. Also $\varphi_U(b'_j) = \sum_k \tau^k_j a_k$. Thus the relationship between w and τ is an elementary question in linear algebra. The result without proof is given in

Lemma 11.7. $\tau = i(I + w)(I - w)^{-1}$, and $w = (\tau - 1I)(\tau + iI)^{-1}$.

The set E of all possible w is the space of symmetric matrices w such that $I - w\bar{w}$ is positive definite. The details may be found in Siegel's book [7] on page 8. The main point of this is that E is a bounded domain. Thus we have biholomorphic identification $\mathrm{Reas}(L \otimes_{\mathbb{Z}} \mathbb{R}) \approx S_g \approx M \approx E$.

§ 11.3 The Jacobian of a Riemann Surface

Let C be a compact Riemann surface ($\mathbb{C}$-analytic curve) of genus g. Thus $H^1(\mathbb{C}, \mathbb{Z}) \approx \mathbb{Z}$ as C is oriented. Therefore the cup-product $\cap : H^1(C, \mathbb{Z}) \times H^1(C, \mathbb{Z}) \to H^2(C, \mathbb{Z})$ define a skew-symmetric integral-valued form E on $H^1(\mathbb{C}, \mathbb{Z}) = L$.

If $[\omega]$ denote the cohomology class of a closed one-form ω on C, then $E([\omega_1], [\omega_2]) = \int_{\mathbb{C}} \omega_1 \wedge \omega_2$. Recall that the space of abelian differentials $\Gamma(C, \Omega_C)$ is a g-dimensional complex subspace and $[\ ]$ induces an injection of $\Gamma(C, \Omega_C)$ into $H^1(C, \mathbb{C}) = H^1(C, \mathbb{Z}) \otimes_{\mathbb{Z}} \mathbb{C} = L \otimes \mathbb{C}$. Let U be the image of $\Gamma(C, \Omega_C)$.

Proposition 11.8. *U is a Hodge structure on L.*

Proof. If ω_1 and ω_2 are holomorphic differentials $\omega_1 \wedge \omega_2$ is zero. So $E([\omega_1], [\omega_2]) = \int_C \omega_1 \wedge \omega_2 = 0$. So P is isotropic. If ω is non-zero holomorphic differential, then $\frac{1}{i}\omega \wedge \overline{\omega}$ is a non-negative measure on $\mathbb{C}$ which is zero at only a finite number of points. Thus $\frac{1}{i}E([\omega] \wedge \overline{[\omega]}) = \int_C \omega \wedge \overline{\omega} > 0$. Thus we are done. $\qquad\square$

Let $A_1, \ldots, A_g$, $B_1, \ldots, B_g$ be a basis of $H_1(C, \mathbb{Z})$ in standard position. Let $a_1, \ldots, a_g$, $b_1, \ldots, b_g$ be their dual basis in $H^1(C, \mathbb{Z})$. This is a canonical basis of $H^1(C, \mathbb{Z})$. We would like to compute the τ-matrix of this Hodge structure. Let $\omega_1, \ldots, \omega_g$ be abelian integrals normalized such that $\int_{B_k} \omega_j = \delta_j^k$. Then $\tau_k^j \int_{A_j} \omega_k$. The conditions that τ is an element of the Siegel's space S_g are called Riemann's bilinear equations and inequality.

Thus $P(U)$ is a principal polarized abelian variety. Hence we have a canonical isomorphism $A(U) = \mathrm{Pic}^0(P(U))$ and $P(U)$. This abelian variety $P(U)$ is called the Jacobian of C. One must be careful to distinguish between the various equivalent forms of the Jacobian but the autoduality of the Jacobian is an important principle.

Next I want to discuss the most direct relationship between the abelian varieties $P(U)$ and $A(U)$ and the curve C. First of all $P(U)$ is the Picard variety $\mathrm{Pic}^0(C)$ of C; i.e. $\mathrm{Pic}^0(C) = \mathrm{kernel}(H^1(C, \mathcal{O}_C^*) \to H^2(C, \mathbb{Z}))$ where $\mathcal{O}$ is the boundary in the exact sequence $0 \to \mathbb{Z}_C \to \mathcal{O}_C \xrightarrow{e^{2\pi i -}} \mathcal{O}_C^* \to 0$. This follows cohomologically as follows. We have an exact sequence $0 \to H^1(C, \mathbb{Z}) \xrightarrow{\alpha} H^1(C, \mathcal{O}_X) \to \mathrm{Pic}^0(C) \to 0$ by diagram chasing. Now $H^1(C, \mathcal{O}_X)$ is represented by the periods of antiholomorphic differentials $\overline{P}$ and α corresponds to the projection of L on this subspace.

For $A(P)$ it is naturally the Albanese variety of C. The complex torus X with a universal analytic mapping $S : C \to X$. By Section 9.1 to define S we need to give a $\mathbb{C}$-linear mapping $R : \mathrm{Hom}_{\mathbb{C}}(\Gamma(C, \Omega), \mathbb{C})^* \to \Gamma(C, \Omega)$ such that for all closed paths γ in C, then linear functional x in $\mathrm{Hom}(\Gamma(C, \Omega, \mathbb{C}))^*$ defined by $\int_\gamma R(\overline{\omega}) = \langle x, \overline{\omega} \rangle$ in $L^\perp$; i.e. $\langle x, \omega_1 \rangle + \overline{\langle x, \omega_2 \rangle}$ in integral if (ω_1, ω_2) is in $H^1(C, \mathbb{Z})$. This is just $\int_\gamma \omega_1 + \overline{\omega}_2 = \langle \gamma, \omega_1 + \overline{\omega}_2 \rangle$. Thus $L^\perp$ is the smallest lattice which satisfies the condition. Therefore we have an integral $\int : C \to X$ and it is clearly universal.

§ 11.4 Picard and Albanese Varieties for a Kähler Manifold

Let Y be a compact Kähler manifold with Kähler form ω. By Hodge Theorem $H^0(X, \Omega_X) \oplus \overline{H^0(X, \Omega_X)} = H^1(X, \mathbb{Z}) \otimes_{\mathbb{Z}} \mathbb{C}$. Thus we have an elementary structure $H^0(X, \Omega_X)$ on $H^1(X, \mathbb{Z})/(\text{torsion})$. Let $\mathrm{Pic}^0(X)$ be P of this Hodge structure and $\mathrm{Alb}(X)$ be A of it. Then $\mathrm{Pic}^0(X) = \{\delta : H^1(X, \mathcal{O}_X^*) \to H^2(X, \mathbb{Z})\}$ as before as $H^1(X, \mathcal{O}_X) = \overline{H^0(X, \Omega_X)}$. Similarly with the one dimensional case we have a universal integral $\int : X \to \mathrm{Alb}(X)$.

One may ask when these tori are abelian. There is a skew-symmetric form $\widetilde{E}$ on $H^1(X, \mathbb{Z}) \otimes_{\mathbb{Z}} \mathbb{C}$ given by $[\sigma_1 \wedge \sigma_2 \wedge [\omega]^{\dim X - 1}]_X$. By local calculation we have $H^0(X, \Omega_X)$ is isotropic and $\frac{1}{i} E[\sigma, \bar\sigma]$ is positive definite of $H^0(X, \Omega_X)$. Thus we have all the conditions but that $\widetilde{E}$ is integral on $H^1(X, \mathbb{Z})/(\text{torsion})$.

Now Y is projective variety and ω represents first Chern class of an ample line bundle. Then $\overline{\omega}$ and hence $\widetilde{E}$ is integral. Therefore we have proven

Proposition 11.9. *The Picard and Albanese of a smooth projective variety are a dual pair of abelian varieties.*

Informal Discussions of Immediate Sources

The statement of the Appel-Humbert Theorem 1.5 is due to Mumford [3]. Also I follow his idea for the proof of the existence Theorem 2.2. Some of the material in Chap. 1 is to introduce the reader to the abstract statements used since A. Weil in the algebraic case. The Sect. 3.4 is modeled on H. Langes lectures notes where he uses ideas from Umemura [8]. The idea of Chap. 4 comes from Mumford's famous paper [4]. Lemma 4.6 is founded in [5]. The Theorem 5.9 and its application in Chap. 8 to the functional equation of the theta function can be found in Igusa [1]. The algebraic material is Chap. 6 was initiated by Mumford [4], added to by Koizumi and Sekiguchi and completed in [2]. Section 8.3 is close to Mumford's discussion in [6]. Proposition 9.7 is due to Serre. Proposition 9.10 is due to Z. Ran. Section 10.1 to 10.3 are an adaptation of H. Lange's notes again. He attributes Theorem 10.1 to Obuki. The Theorem 10.6 is due to Sasaki by a different method. Lange and Narasimhan have informed me that they have another proof. The point of view of Chap. 11 is from that of Deligne's idea of a Hodge structure.

References

[1] J.-I. Igusa: Theta Functions. Springer, New York 1972

[2] G. Kempf: Projective Coordinate Rings of Abelian Varieties. Algebraic Analysis, Geometry and Number Theory. Edited by J.I. Igusa, Johns Hopkins Press 1989, pp. 225–236

[3] D. Mumford: Abelian Varieties. Oxford University Press, Oxford 1970

[4] D. Mumford: On the Equations Defining Abelian Varieties. Invent. math. *1* (1966) 287–354

[5] D. Mumford: Varieties Defined by Quadratic Equations. In: Questions on Algebraic Varieties. Centro. Intern. Mate., Estrivo, Roma, 1970, pp. 31–100

[6] D. Mumford: Tata Lectures on Theta I. Prog. in Math., Vol. 28. Birkhäuser, Boston 1983

[7] C.L. Siegel: Symplectic Geometry. Academic Press, New York 1964

[8] H. Umemura: Nagoya Math. J. *52* (1983) 97–128

Subject Index

S. Gallot, D. Hulin, Ecole Polytechnique, Palaiseau;
J. Lafontaine, University of Montpellier

Riemannian Geometry

2nd ed. 1990. XIII, 284 pp. 95 figs. (Universitext)
Softcover DM 48,– ISBN 3-540-52401-0

This book, based on a graduate course on Riemannian geometry and analysis on manifolds given in Paris, covers the topics of differential manifolds, Riemannian metrics, connections, geodesics and curvature, with special emphasis on the intrinsic features of the subject.

Classical results on the relations between curvature and topology are treated in detail. The book is quite self-contained, assuming of the reader only a knowledge of differential calculus in Euclidean space. It contains examples which are picked up repeatedly to illustrate each new definition or property introduced.

This book addresses both the graduate student wanting to learn Riemannian geometry, and also the professional mathematician from a neighbouring field who needs information about these ideas and techniques which are now pervading many parts of mathematics.

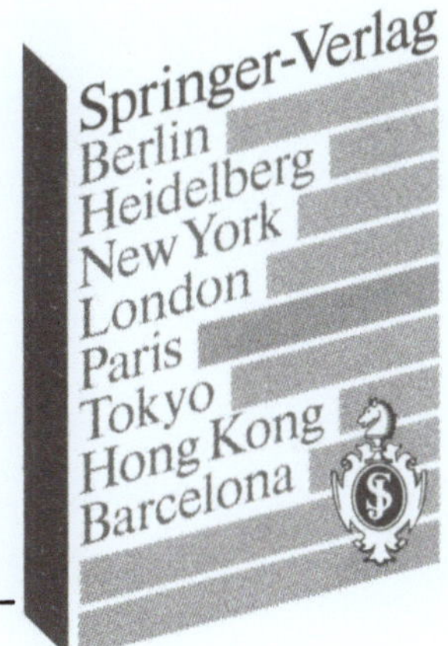